FOOD SCIENCE AND TECHNOLOGY

PROCESSED FOODS

QUALITY, SAFETY CHARACTERISTICS AND HEALTH IMPLICATIONS

FOOD SCIENCE AND TECHNOLOGY

Additional books in this series can be found on Nova's website under the Series tab.

Additional e-books in this series can be found on Nova's website under the e-book tab.

FOOD SCIENCE AND TECHNOLOGY

PROCESSED FOODS

QUALITY, SAFETY CHARACTERISTICS AND HEALTH IMPLICATIONS

CHLOE M. GAGNE
AND
DANIEL B. JONES
EDITORS

nova publishers
New York

Library of Congress Cataloging-in-Publication Data

Processed foods : quality, safety characteristics and health implications / editors, Chloe M. Gagne and Daniel B. Jones.
pages cm
Includes bibliographical references and index.
ISBN: 978-1-62808-688-1 (hardcover)
1. Processed foods--Quality control. 2. Processed foods--Safety measures. 3. Processed foods--Health aspects. 4. Food additives. I. Gagne, Chloe M. II. Jones, Daniel B.
TP370.5.P75 2013
664'.06--dc23

2013029200

Published by Nova Science Publishers, Inc. † New York

Contents

PREFACE

In this book, the authors present current research in the study of the quality, safety characteristics and health implications of processed foods. Topics discussed include novel thermal and non-thermal food processing technologies; persistence of transgenic DNA in processed foods of animal origin; cassava as a source of chemically safe food; transformation of vegetable waste from the food industry into value added products; functional mathematical index (FMI) as an index generator for taming quality as applied to food and processes; nutritional and biological potential of functional foods containing hard-to-cook bean protein hydrolysate; basic guidelines for establishing food standards; strategies for annatto seeds processing with pressurized fluids in the food industry; principles of quality analysis and critical control points applicable to food processing; and engineered colloidal systems from food.

Chapter 1 - Consumers' demand for increased quality characteristics of foods has led to the development of new and less aggressive processing technologies, which allow greater retention of food natural features. Thermal processing operations, such as blanching, pasteurization and sterilization are commonly applied in the food industries in order to attain safety standards of the products. However, these traditional heat-based methods may lead to undesirable sensorial and nutritional changes, such as color degradation, softening of tissues, vitamin losses and development of unpleasant cooked flavors. Aiming at preservation, low temperature processes such as chilling and freezing also inevitably affect quality characteristics.

Alternatively, non-thermal methods, or less severe thermal methods, have been replacing traditional processing technologies. They constitute challenging processes aiming at reducing pernicious effects of thermal methods, while preserving quality and nutritional attributes of foods, and yielding safe and less-perishable product with extended shelf-life.

This chapter provides an overview of traditional and novel thermal and non-thermal processing technologies such as ohmic and dielectric heating, high pressure processes, ultrasounds, UV-C radiation and ozone based treatments, which reveal promising applications in food industry. The impact of those treatments on food safety aspects from a microbiological point of view and on quality characteristics will be discussed.

Chapter 2 - The consumer awareness on the use of genetically modified organisms (GMOs) in food production is growing rapidly, as well as the demand for total transparency.

The authorization of GMOs raw materials for further processing into animal feed or food is subject to strict European standards, and the labeling of their presence just above 0.9% is mandatory.

But there is no obligation to inform the consumer about their use in the production process. It follows that the consumer may be informed of the presence of GMOs in food only when it exceeds the threshold value of 0.9%, while being completely unaware if animal products were obtained by feeding the animals with even 100% GMOs.

Processed foods have gained popularity in recent years, many of the most produced being of animal origin, such as dairy products, meat and fish.

One of the most studied GM food safety concern is the fate of ingested recombinant DNA throughout the food chain, from GM crops through the gastrointestinal tract (GIT) of livestock, then to animal tissues and offspring and subsequently to humans. The main biological risk is the occurrence of horizontal transgenic transfer (HGT) to bacteria in the host GIT or animal cells, with many, almost unexplored implications.

The goal of this review is to organize and summarize the significant amount of research regarding the fate of dietary DNA in livestock and its persistence in animal origin processed food. The main technique of investigation is Polymerase Chain Reaction (PCR). Many results of recent publications agree that most of the DNA is degraded in the GIT by endogenous restriction nucleases, nevertheless some DNA fragments have been found in animal tissues and in products such as milk and meat.

Many discordant results have been reported about the presence of transgenic DNA, due to methodological limitations. Therefore additional research is needed in the field of both genetic traceability studies and DNA quantification methods development.

Chapter 3 - The issue of food contaminants is broad ranging from undesirable natural constituents or unauthorized additives over environmental pollutants and to contaminants related to specific steps of food production, such as fermentation, heat treatments or to cross-contamination. In some cases, mainly mycotoxins and toxic natural plant constituents and their derivatives pose a risk for animal and human welfare. Risk assessment issues include the identification and characterization of the toxicological hazards, identification of the possible routes of the contaminants to plant based food and the carry-over of the parent compounds or their metabolites to animal derived foods; all in order to define exposure scenario(s). Based on this situations that may require risk management measures can be pinpointed. Some scientists, regulators, non-governmental organizations and consumers are raising concerns about potential risks related to the increasing use of cassava and its products.

In this paper, the authors broadly review risks due to chemicals that may be present in cassava (*Manihot esulenta* Crantz) products and provide a brief overview of some of the acceptable limits comparing levels in cassava and other crops. In conclusion, the authors point out the safety of most cassava products.

Chapter 4 - Byproducts of vegetable food processing represent a major disposal problem for the industry. Transformation of vegetable waste of the food industry into value added products as fibres are, may contribute to diminish pollution and to recover valuable biomass and nutrients.

A great number of diseases frequently occurring in developed countries such as various gastrointestinal disorders (hernia, duodenal ulcer, gall stones, Crohn's disease, appendicitis, constipation, hemorrhoids, colon carcinoma), diabetes mellitus, obesity and cardiovascular diseases, have a lower incidence among people consuming high amounts of fibre. The benefits of the intake of fibres for the gut are amply recognized, fact that has determined its consideration as a nutrient. Fibre addition to foods is an alternative to compensate for the existent deficiency in the diet.

Besides the nutritional aspect, fibres can be used for technological purposes because they can change food consistency, texture, rheological behavior and sensory characteristics.

As only the tender inner bracts and the receptacle of the immature flowers of globe artichoke (*Cynara scolymus* L.) are used for consumption, residues can be upgraded to add value to its production, for example through their use for dietary fibre isolation.

In this research, it was evaluated a procedure for the obtention of fractions enriched in soluble fibre from artichoke stems. These tissues were dehydrated (80°C–2.5 h), milled, and dispersed in citrate buffer. Afterwards, cellulase was added and suspensions were incubated (30°C–20 h). Procedures were repeated but without enzyme. Soluble fibre fractions were precipitated with ethanol, filtrated, washed and freeze-dried. Yields obtained were 5% in the absence of enzymes and 7% when cellulase was used.

Fractions obtained were completely characterized observing a 69-70% (w/w) content of carbohydrates and a 25-28% (w/w) content of proteins. Carbohydrates included uronic acid and inulin. The use of the enzyme not only produced a yield increase but also an increase in uronic acid content.

Both fractions helped to stabilize oil-in-water emulsions probably due to their high protein content. The flow behavior of aqueous solutions of both fractions was pseudoplastic. The use of cellulase for fraction obtention, determined a lower consistency index and the existence of a minimum shear stress.

Procedure proposed allowed to isolate fractions enriched in soluble fibre, adding value to left over of *Cynara scolymus* industrialization, helping in the development of green processes for the agrifood chain and providing natural ingredients for more healthy foods.

Chapter 5 - In several fields, e.g. in Nutritional Sciences and Food Technology, it is important to establish the right way to detect and define the quality of food and related processes.

Generally speaking the quality depends not only on the choice of measurable parameters, allowing to determine food quality in an objective way, but also on non “scientific” and nonmeasurable aspects.

In a general way useful parameters involved in quality determination are not comparable to each other (i.e. different measure units and scales). In addition, in literature several quality measurement techniques make use of weights - associated to the characteristic parameters - which introduce a sort of subjectivity in defining quality.

In order to solve these methodological problems, for their purposes the authors developed a Functional Mathematical Index (FMI) that allows to define the quality in food and processes without weights. FMI is a number obtained by a suitable choice of a set of parameters that in a unique way characterizes the quality of food and processes. The choice of number and type of parameters allows to make some particular features to raise up.

Mathematically speaking, FMI was introduced as the norm of a vector in an N-dimensional space, where N is the number of chosen parameters. The chosen parameters are normalized (their values are between 0 and 1), and summed together. Further studies of the authors’ research group have modified and improved the mathematical meaning of FMI, which however can be interpreted, in a very general way, as the distance from the “ideal” food and the considered sample. The better the quality is, the closer to 0 FMI is. The core of FMI building process is the appropriate choice of the parameters.

We tested FMI with good results on different products, e.g. extra virgin olive oil along the ripening phase, black and green tea, several potato cultivars cooked in many ways, durum wheat etc.

The FMI shows a wide ductility and adaptability to a wide range of foods and to different application fields.

In order to evaluate the importance and the usefulness of FMI, in this review the authors will summarize the development of this index and the authors will show the last definition applied onto food and technical processes, with numerical examples.

Chapter 6 - Advances in understanding of the relationship between nutrition and health contributed to development of the concept of functional foods, a practical and new approach to achieving optimal health through promoting well-being and possibly reducing disease risk. Functional foods are found in virtually all food categories. Pasta is considered a functional food for its low fat and sodium contents. Wheat semolina is highly suitable for making pasta because its proteins are ideal for dough development and for preventing pasta disintegration during cooking. However, wheat pasta is nutritionally unbalanced due to its low protein content (<15%) and relatively deficient levels of the essential amino acid lysine. Wheat pasta is an effective vehicle for food supplementation with minerals, proteins and other health-promoting components. The present chapter objective was to evaluate the nutritional and biological potential of semolina pasta added with hard-to-cook bean (*Phaseolus vulgaris* L.) protein hydrolysate. Hard-to-cook bean protein hydrolysate obtained with the Alcalase®-Flavourzyme® sequential enzymatic system was added to durum wheat semolina pasta at two concentrations (5 and 10%). The pastas exhibiting angiotensin I-converting enzyme (ACE) inhibitory activity (10.88-17%) and antioxidant activity (TEAC = 26.27-31.42 mM/mg protein). Evaluated nutritional parameters and sensory scores indicated the best replacement level to be 10% hydrolysate. Addition of a hard-to-cook bean hydrolysate to semolina pasta produced by extrusion improved product nutritional value without significantly affecting sensory parameters. Semolina pasta containing 10% protein hydrolysate from hard-to-cook *P. vulgaris* can be considered a functional food with both biological activity and a good nutritional profile.

Chapter 7 - Food manufacturers need a previously established standard to produce high quality food products with uniformity, which is always concerned by consumers. This article aims to introduce the basic guidelines that should be followed in developing a common accepted food (especially new food) standard. Based on the literature review and evaluation of the current status of food quality standards, the basic guidelines for establishing a food standard have been developed. The contents of this guideline include *Determining the critical interior characteristics of a food product required by its standard, Determining the quality requirements of raw materials and additives for producing a food product by its standard, Determining the critical foreign material contamination of a food product required by its standard, Determining the required processing, storing or selling conditions of a food product by its standard, Determining the critical limit level of the required characteristics of the food product by its standard, Establishing the framework of the standard* and *Drafting the standard according to the framework established.* The publication of this guideline should be able to help any company or agency to establish a well common accepted standard for a well established or newly developed food product.

Chapter 8 - Nowadays, food scientists have collaborated with nutrition researchers to develop plant-based food products to promote healthy eating habits. In food research, carotenoids from fruits and vegetables have attracted a great deal of attention for their functional properties, health benefits and prevention of several major chronic diseases. Annatto (*Bixan orellana L.*) is a tropical tree whose seeds produce carotenoid pigments, oil-soluble bixin and water-soluble norbixin, which have a widespread use in the food industry, namely for coloring butter, margarine, cheese, oils and sauces, with hues ranging from yellow to red. Carotenoids are unstable compounds, very sensitive to oxidation processes. To avoid these effects some caution during its processing should be taken into consideration. The processing of carotenoids with pressurized fluids in liquid, subcritical or supercritical state at selected conditions has demonstrated to be a promising option. This chapter presents the recent works done and under evaluation by our research group (LASEFI/ DEA/FEA/UNICAMP) for obtaining annatto seeds products: extracts, coprecipitates, nanosuspensions and nanoemulsions aiming higher yields, low manufactured cost with concomitant minimal carotenoid degradation.

Chapter 9 - Consumer's choice of a food product is usually based on its good quality. Food manufacturers always produce a food product that is demanded by consumers. The target of QACCP system is to assure the processing of the food product with its quality as high as possible and its quality controlling cost as low as possible. This would attract food manufacturers. Implementation of QACCP follows eight principles including *Principle 1- Establishment of the principles for quality analysis*, *Principle 2-Determination of critical factors according to food standard established*, *Principle 3-Recognition of critical control points by quality analysis*, *Principle 4-Establishment of critical limits for each critical control point*, *Principle 5-Creation of measures suitable for monitoring the status of critical limit for each critical control point*, *Principle 6-Creation of procedures for corrective action*, *Principle 7-Creation of procedures for verifying the QACCP system* and *Principle 8-Creation of procedures for keeping record or memorandum document*. By strictly following the 8 principles of QACCP described above, it is reasonable to believe that any food product processed can at least meet its standard previously established. It is believed that the tables provided are well designed, quite accurate and easily operative. They should be very helpful for putting the 8 principles of QACCP into practice. Therefore, the application of QACCP to food manufacturing may be very perspective. The 8 principles of QACCP may also be helpful for food scientists and engineers. They may be included in a text book to teach students in food quality control.

Chapter 10 - Though the novel processed foods have been a topic of global interest in the food and nutrition industries for years, the general consumers seemed incognizant of the obvious benefits that these novel foods could offer. However, recently with the amount of information available in the public domain, modern consumers are becoming more health conscious about their food choices and consequently the market for high-quality novel food products is constantly evolving. As a result, the demands on the improvement of existing food products and the development of novel ones are increasing to such an extent that food industries globally are feeling the necessity to invest heavily into inter-disciplinary research. One of the most important factors to be considered while working in the foods domain, is the choice of raw materials. The strict regulatory requirements in foods means that the starting materials used for the fabrication of novel formulation and delivery systems are food-grade and edible. This chapter gives an account of two such food-grade materials (zein, a hydrophobic plant protein) and (shellac, a hydrophobic resin), which have been extensively utilized as materials of choice to fabricate different types of colloidal delivery systems. Specifically, the chapter focuses on an overview of a range of engineered colloidal systems such as colloidal particles, responsive emulsions and microcapsules prepared using zein and shellac.

In: Processed Foods
Editors: Chloe M. Gagne and Daniel B. Jones
ISBN: 978-1-62808-688-1

Chapter 1

NOVEL THERMAL AND NON-THERMAL FOOD PROCESSING TECHNOLOGIES

Elisabete M. C. Alexandre, Teresa R. S. Brandão and Cristina L. M. Silva

CBQF, Centro de Biotecnologia e Química Fina, Laboratório Associado, Escola Superior de Biotecnologia, Universidade Católica Portuguesa/Porto, Portugal

ABSTRACT

Consumers' demand for increased quality characteristics of foods has led to the development of new and less aggressive processing technologies, which allow greater retention of food natural features. Thermal processing operations, such as blanching, pasteurization and sterilization are commonly applied in the food industries in order to attain safety standards of the products. However, these traditional heat-based methods may lead to undesirable sensorial and nutritional changes, such as color degradation, softening of tissues, vitamin losses and development of unpleasant cooked flavors. Aiming at preservation, low temperature processes such as chilling and freezing also inevitably affect quality characteristics.

Alternatively, non-thermal methods, or less severe thermal methods, have been replacing traditional processing technologies. They constitute challenging processes aiming at reducing pernicious effects of thermal methods, while preserving quality and nutritional attributes of foods, and yielding safe and less-perishable product with extended shelf-life.

This chapter provides an overview of traditional and novel thermal and non-thermal processing technologies such as ohmic and dielectric heating, high pressure processes, ultrasounds, UV-C radiation and ozone based treatments, which reveal promising applications in food industry. The impact of those treatments on food safety aspects from a microbiological point of view and on quality characteristics will be discussed.

1. INTRODUCTION

Food processing may be generally defined as the conversion of raw materials or ingredients into a food product, or the transformation of food into other forms that are more convenient or more appellative to be consumed, aiming at extending shelf life of the products (Hartel and Heldman 1997; Heldman and Hartel 1999). Initially, food was processed with the purpose of preservation. Later, it was also discovered that food taste, texture and appearance can be attractively changed by applying different processes. Nowadays, the main objectives of food industries can be described as: *(i)* extension of foods shelf life, which is usually attained by the use of preservation methods that inhibits microbiological and biochemical changes, allowing a safe distribution, sale and home storage; *(ii)* improvement of organoleptic characteristics, such as attractive flavors, colors, aromas, textures, shapes by changing the original form of foods; *(iii)* guarantee of food nutritional characteristics; and *(iv)* economic interests (Fellows 2000).

Temperature is a key factor in food processing, since high temperatures allow microbial death or inactivation, whereas low temperatures are applied for long term food preservation, preventing microbial growth and retarding reactions of quality alterations, in a joint perspective of safety and quality.

Thermal treatments such as pasteurization, sterilization, drying and evaporation are frequently applied in the food industries in order to attain safe products. However, undesirable sensorial and nutritional changes, such as color degradation, softening of tissues, vitamin losses and development of unpleasant cooked flavours may occur. On the other hand, the use of low temperatures to preserve foods is also frequently applied. Freezing and chilling processes have a strong impact in enzymatic activity and oxidative reactions that help avoiding products deterioration. In addition, with ice crystal formation, less water will be available to support deteriorative reactions and microbial viability (Heldman and Singh 1981; Singh and Heldman 2009). However, chemical changes such as lipid oxidation, protein denaturation, enzymatic browning and degradation of pigments and vitamins may be verified during freezing and frozen storage. The risk of freeze cracking, moisture migration, recrystallization of ice crystals and drip loss during thawing also may happen. Moreover, above $0^{\circ}C$ the loss of microorganisms' viability is limited and in practice, is negligible. All of these alterations reinforced the search of new and gentle processing technologies that prolong shelf-life without the detrimental effects caused by severe temperatures. Some of the most recent advances in food processing operations include ohmic heating, dielectric heating, which includes radio frequency and microwave heating, high pressure, ultrasounds, ozone treatments and UV-C radiation. All these technologies constitute challenging processes aiming at reducing pernicious effects of thermal methods, by preserving quality and nutritional attributes of foods and yielding safe and less-perishable products.

Food quality characteristics and food safety aspects will be following presented, once their preservation are the main purpose of applying processing technologies.

2. Food Quality Characteristics

Food quality is a multidimensional concept that includes chemical, physical and sensory characteristics of the products. Consumers can scrutinize some external quality attributes. Overall appearance, size, shape, aroma, color, state of maturation and defects can be easily evaluated by vision, touch and smell. However, internal quality attributes such as taste and texture can be more difficult to evaluate, because they can only be assessed if consumer eats the product. Moreover, hidden quality characteristics like nutritive value and wholesomeness are impossible to measure at purchase (Anonymous 2002; Rico et al. 2007). Quality parameters may be assessed by sensory inspection or by quantification and instrumental measurements.

Quality is also associated with healthy aspects, which include food components that have a positive and functional impact on human health. The human body has complex nutritional requirements that must be fulfilled. The consumption of fresh fruits and vegetables has a significant benefit to human health. Antioxidants, minerals, vitamins and fibers in produce are good examples of food components that can help in the prevention of diseases. For example, calcium has a significant role in the prevention or control of osteoporosis, folate is associated with the prevention of neural tube defects in infants and fiber is associated with the prevention of colon cancer. Some phytochemicals such as carotenoids present in carrots and the yellow, red, and orange pigments of plants are highly significant in the prevention of cancer, diabetes and hypertension. Vitamin A (found in carrots, spinach and cantaloupe), vitamin B (present in broccoli, strawberries and tomatoes) and vitamin C (found in significant levels in oranges, kiwifruit and mango, and in certain types of nuts) are examples of compounds that can help to minimize cells oxidative damage and have a role in the immune system. Meat and seafood are also important sources of a wide variety of essential nutrients. One of the most important features of meat is its high proteins content, which plays an important role in human nutrition as source of essential amino acids such as leucine, lysine, threonine, methionine and tryptophan required for maintenance and growth of human tissues (Varnam and Sutherland 1995). Fish is perhaps more important as supplier of micronutrients, minerals and essential fatty acids than for its energy or protein values. Vitamins A and D, calcium, phosphorus, magnesium, iron, zinc, selenium, fluorine and iodine are some examples of the essential micronutrients and minerals that are present in fish (Arino et al. 2005).

In the recent years, increased attention has been given to food quality, not only from a consumers' perspective but also from governmental entities. To ensure consumer satisfaction, some food standards were designed to give information about the product and preservation of its quality. Food standards were also developed to prevent economic fraud and to establish market value of products (Anonymous 2002; Keast 2010). There are various institutions worldwide that specialize in the preparation of food standards and codes of practice for food products and for the performance of the quality inspection and certification of food products.

3. Food Safety Aspects

Any single definition of food safety will be overly simplistic because it is a complex and multifaceted concept that results from consumers, special interest groups, regulators, industry, and academia descriptions based on their own perspectives (Skip and Seward 2003). A general definition of food safety is related with the guarantee that food will not cause harm to the consumer when it is prepared and eaten according to its intended use. It can be considered as a degree to which all quality requirements relating specifically to features or food properties that have the potential to be harmful to health or to cause illness or injury are met. For example, counts of total bacteria, coliform bacteria or mold and yeasts can be considered a food quality feature that can be used as indicators of food safety. Thus, when a food product is not conforming to the food safety requirements, inevitably it is not conform to food quality requirements (at least one quality parameter is not verified). If the food is conform to the food safety requirements, it may be or not be conform in terms of food quality, it depends on the other quality requirements.

Several products are commonly involved in food poisoning incidents. Examples may include products so diverse as meat and poultry and their products, seafood and seafood products, egg and egg products, milk and dairy products, fruits and vegetables, low-acid canned foods or even water.

Raw products often contain a great diversity of microbial flora and are frequently involved in food-bourne outbreaks, which may cause serious illnesses. Since fresh products, such as fruits and vegetables, are mainly consumed uncooked or minimally-processed, microbiological safety becomes a very important issue to minimize consumers' risks (Sagoo et al. 2003).

3.1. Biological Hazards

Food-borne pathogenic bacteria are responsible for a large number of food poisoning incidents. There are more than forty different bacteria known as pathogenic, which fit into three categories: *(i)* infectious invasive agents, such as *Salmonella* spp., *Listeria monocytogenes*, *Escherichia coli* and *Campylobacter* spp.; *(ii)* toxigenic pathogens like *Staphylococcus aureus*, *Clostridium botulinum* and *Bacillus cereus*; and *(iii)* toxico-infectious agents such as enterotoxigenic and enterohemorrhagic *Escherichia coli* and *Clostridium perfringens*. The symptoms of contamination usually include headache, muscle pain, nausea, fatigue, chills or fever, stomach or abdominal pain, vomiting, and diarrhea. Children and elderly are particularly vulnerable groups where numerous severe and fatal illnesses may occur as a result of food poisoning due to pathogenic bacteria (Behling et al. 2010; Waite and Yousef 2010).

Some of those pathogens are naturally present in the environment and one of the main sources of contamination is human and/or animal fecal deposits. Pathogens can be present in soils, irrigation and/or washing waters, and in organic fertilizers or soil conditioners derived from waste water sludge; insects and birds can be vehicles of contamination (Anonymous 2001; Warriner 2005; Zhao 2005; Gorny 2006).

Foods can also be agents of transmission of parasites, including parasitic protozoan species such as *Cryptosporidium parvum* and *Entamoebahistolytica* and parasitic worms such as *Taeniasolium* and*Ascarislumbricoides.* The most common virus transmitted by food are the hepatitis A and E viruses, the Norwalk group of viruses and rotavirus (Waite and Yousef 2010).

3.2. Chemical Contamination

Chemical and biochemical contaminants such as heavy metals and persistent organic pollutants (*e.g.* polycyclic aromatic hydrocarbons, PAH) are frequently found in fresh products. They derived from the application of pesticides, insecticides, herbicides, rodenticides, fungicides, fertilizers, drugs, hormones, and antibiotics. Many pesticides, due to their potent neurotoxic chemicals, attack and disable portions of the nervous system. Compounds containing lead, arsenic, copper, mercury and zinc are frequently used on annual crops and fruit trees (Alloway 2004; Waite and Yousef 2010).

Fertilizers, which are also a concern, are broadly classified in two categories: organic (composed by decayed plant or animal matter) and inorganic (composed by simple chemicals and minerals). These materials may contain significant concentrations of trace metals including cadmium and zinc (in phosphate fertilizers) and a wide range of metal and organic contaminants (in composts and sewage sludges) (Alloway 2004).

Industrial contaminants that enter in the environment as a result of industrial activity may include heavy metals (lead, mercury, arsenic) and organo-chlorinated compounds, such as polychlorinated biphenyls (PCBs). Those compounds may contaminate foods through environmental pollution.

The industrial cleaning and sanitizing processes of surfaces and equipment with chemical compounds that are not allowed by regulatory entities, may also compromise food safety.

4. Traditional Processing Methods

The traditional methods aiming at food preservation are based on high temperatures to guarantee the destruction of pathogens and spoilage microorganisms and inactivate enzymes that are responsible for quality alterations. Processes based on low temperatures, such as refrigeration and freezing, slow down deteriorative reactions and prevent microbial growth. One should bear in mind that refrigeration or freezing cannot improve the quality of decayed food; they can only retard deterioration. The higher the quality of the product before storage, the better expected final quality before consumption.

A number of sanitising agents may be used to reduce the risk of contamination. Traditionally, fresh fruits and vegetables are rinsed in chlorine, iodine, hydrogen peroxide and acid solutions. However, the impact of those treatments depends on initial microbial loads and, for high contamination and especially in the case of pathogens, those washing solutions may not be efficient.

4.1. Thermal Treatments

To achieve a safe product with extended shelf life it is necessary to destroy pathogenic and spoilage microorganisms and their spores, and to inactivate enzymes responsible for quality decay. However, heath as a negative impact on organoleptic and physical important characteristics of the products and should be conveniently designed with the purpose of achieving safety and maximum quality retention.

Thermal technologies involve the use of hot air (heat transfer is only moderate, but it can be increased through air humidification), steam (heat transfer is very good due to steam condensation), hot water (heat transfer is relatively good), hot fat or oil (very good heat transfer properties) or extrusion (Varnam and Sutherland 1995). According to the treatment severity, conventional thermal methods can be divided into blanching, pasteurization and sterilization.

Blanching is one important process that precedes other preservation processes such as freezing and dehydration. It is mainly applied to vegetables with the purpose of decreasing enzymes activity, which are responsible for undesirable colors and flavors, texture alteration and nutritional decay. The process is characterized by the use of short bursts of heat lasting about 2-3 minutes, at temperatures usually between 85ºC and 100ºC. The definition of blanching time/temperature to which the product must be submitted depends on the product, its size, shape and thermal properties and enzymes activity (Fennema et al. 1973; Knorr 1995). The negative aspects of blanching are related to the effects of heat on sensorial and nutritional characteristics of foods. Softening of tissues, vitamin loss, development of undesirable cooked flavors and loss of soluble solids (to the blanching water) are some examples (Qi et al. 1995; Martins and Silva 2001).

Pasteurization eliminates pathogenic vegetative cells, however spores and heat resistant microorganisms can survive. To minimize the microbial surviving cell, pasteurization is often combined with other processes, such as refrigeration, chemical additive addition, vacuum packaging and fermentation. Pasteurized products generally reach an internal temperature of 71°C, but time/temperature ratio depends on the food heat resistance and its sensitivity to heating.

Sterilization eliminates or kills all forms of microbial life, including spores and heat resistant microorganisms. A sterile food is therefore one food free from any living entity. Industries apply the concept of *commercial sterility*, which is commonly used in the canning industry, meaning the condition achieved by the application of heat sufficient to render the processed product free from viable microorganisms (including those of known public health significance), capable of growing in the food under normal non-refrigerated temperatures at which the food is likely to be held during distribution and storage. Commercially sterile foods are heated, for varying periods of time, at high temperature (up to 121°C) and *Clostridium botulinum*, *Clostridium sporogens* and *Bacillus stearothermophilus* are frequently used as indicators microorganisms, to assure the success of the sterilization procedure. Sterilized foods have a considerably extended shelf life at any ambient temperature, provided it is kept sealed. However, sterilized products are markedly different from fresh ones. For maximum quality retention and totally destruction of spoilage and toxigenic microorganisms, it is convenient to use short exposition times and high temperatures (Lawrie 1985; Potter and Hotchkiss 1998).

4.2. Low Temperature Preservation

Chilling is the most widely used and effective mean of preserving fresh foods for short periods, by lowering the product temperature above its freezing point. Usually, commercial and household refrigerators operate between 4.5 and 7°C, but some commercial refrigerators operate at slightly lower temperatures. Pure water freezes at 0ºC, but most foods do not freeze if -2ºC or lower temperatures were not attained (Potter and Hotchkiss 1998; North and Lovatt 2012).

Chilling and cold storage has relatively few adverse impacts on food quality attributes, such as taste, texture and nutritive content. The products surface temperature should be lowered as fast as possible, to prevent microbial proliferation and to minimize the rates of quality decay reactions. The most important factors that influence the chilling rate of a product are the air temperature and the air velocity. Relative air humidity affects mass losses; decreases of relative humidity implies higher mass losses (Potter and Hotchkiss 1998; Sotopforth and Sofos 2005).

Freezing is one of the oldest and most frequently used processes for long term food preservation in which the temperature of a product is reduced below its freezing point. This preservation method may be used for all kind of commercialized foods: fruits (whole, puréed or as juice concentrates) and vegetables; fish fillets and seafood; meats and meat products; baked goods (as bread, cakes, pizzas); desserts and an endless number of pre-cooked dishes. However, changes in the structure of the product occur due to ice crystals formation when energy is removed (Fellows 2000).

Food preservation by freezing occurs through different mechanisms. When temperature is lowered below 0°C, there is a reduction in the microbial loads and microbial activity; therefore, deterioration rates of foods decrease. Freezing temperatures affect biological materials in various ways, depending on their chemical composition, microstructure and physical properties. Low temperatures have also a strong impact on enzymatic activity and oxidative reactions that help avoiding products deterioration. In addition, with ice crystal formation, less water will be available to support deteriorative reactions and microbial viability (Heldman and Singh 1981; Singh and Heldman 2009).

The characteristics of the final frozen products are influenced by processing factors such as the freezing method, freezing rate, final temperature and frozen storage conditions. Quality attributes such as texture, color, flavor and nutritional content are mainly related to the way ice crystals are formed. Usually, when a plant or animal tissue is cooled, the ice crystals are initially formed at the products surface and their growth depends strongly on the freezing rate. Some products require fast freezing rates (short freezing times) to assure the formation of many but small ice crystals within the product structure, avoiding cells shrinkage and reducing to minimum the degree of freeze damage. In this case, texture changes and small loss of nutrients will be verified through drip on thawing. Additionally, a rapid freezing followed by a slow thawing inactivate more microorganisms. Some products, due to their geometric configurations and sizes, do not allow a fast freezing. If the product is cooled slowly (high freezing time), large ice crystals will be formed, causing maximum disruption of tissues structure. The frozen storage temperatures have also a strong influence on products characteristics. Fluctuations and abuses of the storage temperature may be harmful to products quality (Kennedy 2003; Singh and Heldman 2009; Tucker 2011).

4.3. Chemical Treatments

A number of sanitizing agents may be used to reduce the risk of contamination. Traditionally, several fresh foods are rinsed in chlorine, iodine, hydrogen peroxide and acid solutions.

Chlorine is well known by its impact on microbial loads and was one of the first chemical treatments implemented in industry. It has been used as an effective disinfectant of drinking water, wastewater, and food processing equipment and surfaces. The chlorination process usually consists in adding chlorine by addition of sodium or calcium hypochlorite to wash waters. It has been reported that chlorine has a bacteriostatic effect. However, the lack of effectiveness of chlorine solutions in fresh foods disinfection it is correlated to the neutralization of chlorine by tissues components, thus limiting its oxidative action to microorganisms (Beuchat 1998; Bachmann and Earles 2000; Acuff 2005; Alexandre et al. 2012).

Iodine is used to sanitize food processing equipment and surfaces. Iodine formulations, called iodophors, are water-soluble, less volatile and less irritating to the skin than other iodine formulations. Iodophors inactivate a wide spectrum of microorganisms. At low temperatures are less corrosive than chlorine and are minimally affected by the presence of organic compounds. However, at higher temperatures (above 50°C), they vaporize easily and become very corrosive. The application of iodophors to food products is restricted, as iodine forms a purple complex with starch (Beuchat 1998).

Hydrogen peroxide is another well-studied oxidizing agent that is toxic to microorganisms. Hydrogen peroxide has a very low toxicity to humans and it is recognized as ecofriendly due to its low impact on environment. However, considering different food types, it has been reported that hydrogen peroxide can negatively affect the quality of some products. Hydrogen peroxide may induce browning (mushrooms, lettuce), may cause bleaching of anthocyanins (strawberries and raspberries) and it is not efficient in killing yeasts and moulds (Sapers and Simmons 1998; Bachmann and Earles 2000).

Organic acids such as acetic, citric, succinic, malic, tartaric, benzoic and sorbic acids are commonly used in fresh products disinfection. Pathogenic microorganisms present in products surfaces are killed or inhibited through washes and sprays with such materials.

The effectiveness of these chemical solutions depends on initial microbial loads and microbial sensitivity to the sanitizer agent used. In products with high bioloads, such treatments may not be sufficient. In part, the lack of efficiency can be attributed to the inaccessibility of locations that support microbial flora.

5. Novel Thermal Food Processes

In conventional heating processes, the food internal resistance due to heat conduction can result in heterogeneous heat treatment, leading to notable loss of textural and microbiological quality of the products. Novel thermal processing technologies, such as ohmicand dielectric heating based ones, overcome such limitations and have been gaining increased industrial interests (Yildiz-Turp et al. 2013). Moreover, these technologies are cleaner, more environment friendly and energy efficient than conventional methods currently in use.

5.1. Ohmic Heating

5.1.1. Basic Principles

Ohmic heating (OH) is also known as Joule heating, electric resistance heating, direct electric resistance heating, electroheating and electroconductive heating. It is a process in which alternating electric energy is passed directly through a liquid or solid food product, resulting in an instantaneously generation of heat within the product due to the electrical resistance of the food (Lima 1999; Pereira and Vicente 2010). It is necessary a direct contact between the electrodes and the food product to be heated, once OH relies on direct ohmic conduction losses in a medium (Figure 1).

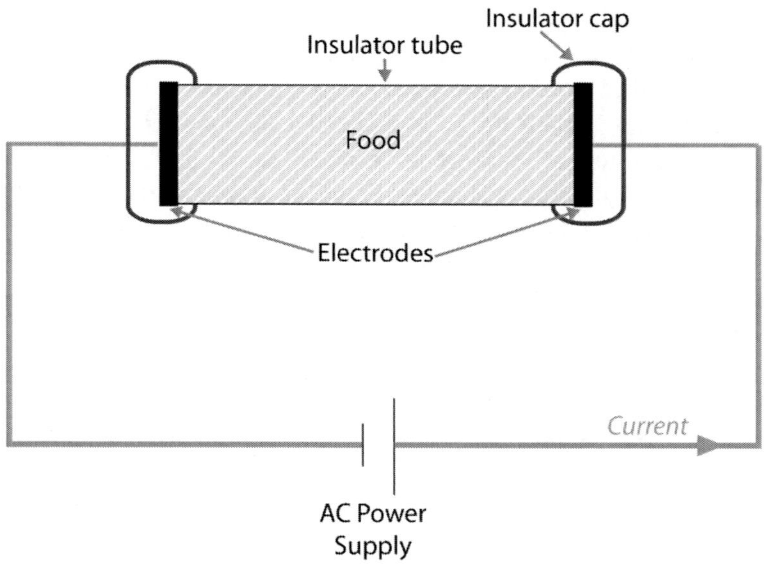

Figure 1. Schematic illustration of an ohmic heating process.

The basic principle of OH is the dissipation of electrical energy in the form of heat through an electrical conductor. The energy generation rate can be calculated by:

$$G = E^2\sigma \quad \text{(Equation 1)}$$

The heat power per unit volume (G) is proportional to the square of the electric field strength (E) and electrical conductivity (σ), which is a function of temperature, type of material and the heating process (Goullieux and Pain 2005; Sastry 2005; Pereira and Vicente 2010). Generally in solid foods σ increases with temperature and the relationship between σ and temperature becomes linear with increasing field strength values. In liquid foods, σ always presents a linear relationship with temperature if non-conductive constituents are absent. Therefore, in particulate fluids, particles heat up faster than the transport fluid, depending on σ values. To improve the effectiveness of heating in those situations, the ionic content of products to be heated can be altered by the addition of electrolytes, such as NaCl. With similar values of σ for all the components, the product heats uniformly (Pereira and Vicente 2010). Thus, the success of the OH treatment depends essentially on heat generation, mainly related to the electrical conductivity and residence time distribution (Goullieux and Pain 2005; Sastry 2005). Other parameters such as particle orientation, geometry, size and concentration may also influence the heating rate of the different constituents of a food (Pereira and Vicente 2010).

5.1.2. Impact on Cellular Structures

The main mechanism of microbial inactivation seems to be related with temperature. However, some reports suggest that other mechanisms, such as a mild electroporation may occur, leading to non-thermal cellular damages. This additional effect may be due to the low frequencies (50–60 Hz) used in OH processes. The specific dielectric strength of cell membranes can be exceeded due to the exposure to electric field unrolling the electroporation mechanism, which leads cell walls to build up charges and form pores (FDA-CFSAN 2000). Depending on the strength of the electric field, the pores formed can vary in size and can reseal after a short period of time. Excessive exposure to the electric fields causes the leakage of intracellular components through the pores and cell death may occur. Thus, electroporation process is greatly harmful to a cell and would increase the lethal effects of the thermal abuse already imposed by OH (Aymerich et al. 2008; Zell et al. 2010;Yildiz-Turp et al. 2013).

Ohmic heating allows faster inactivation rates of peroxidase, pectin esterase, lipoxygenase and polyphenoloxidase than conventional heating processes. Flavor compounds are not degraded so quickly as during conventional pasteurization, also due to shorter residence times (Jakób et al. 2010).

Although there are limited researches about the impact of OH on microbial contamination of specific products, it was found that this method is an efficient process for enzymes and microorganisms inactivation, especially in liquid foods (Yildiz-Turp et al. 2013).

5.1.3. Principal Advantages

The major advantage of OH is related with the uniform heat generation that results in a uniform temperature distribution (volumetric heating). This avoids the usual heat damage associated with the excessive surface heating of conventional thermal treatments and some non-uniformity commonly associated with microwaves heating due to limited dielectric penetration. As OH accompanies the current, the heat distribution throughout the product is also faster than in conventional thermal methods. Consequently, shorter processing times are required and higher yields are obtained, resulting in a safe product with better sensorial and nutritional characteristics. This process is energetically efficient, once 90% of the energy can be converted in heat (Lima 1999; Ohlsson 2002; McKenna et al. 2006; Yildiz-Turp et al. 2013).

5.1.4. Food Industry Applications

Ohmic heating has been applied to a wide variety of foods, including liquids, solids, and liquid–solid mixtures. One of the first products to be processed with OH was milk in the 1930s. Later, in 1980, the process was revived and applied in the United States to process liquid egg and in the United Kingdom and Japan to process whole fruits, such as strawberries (Lima 1999; Sastry 2005). Today, some of commercial applications pass through aseptic processing of high-added-value ready-meals stored and distributed at ambient or refrigerated temperatures, pasteurization of particulate food products and pre-heating of foods before canning (Fellows 2000; Ohlsson 2002). However, this process is mainly considered as an alternative heating system to cook and sterilize liquid food and solid–liquid mixtures (*e.g.* pumpable foods) in continuous flow mode (Zell et al. 2009). The use of OH to process solid foods is limited to comminuted materials (Zell et al. 2010; Yildiz-Turp et al. 2013). One of the greatest limitations to implement this technology in solid foods, such as meat, is related to the electrical nature of muscle compounds. Compounds with poor conductivity, such as fat, do not generate heat at the same rate as muscle and, consequently, a cold spot can be created. Moreover, characteristics like muscle fiber direction, size of meat piece and the type of meat affect the electrical conductivity, which also limits the use of OH in solid meats (Aymerich et al. 2008; Yildiz-Turp et al. 2013).

The limitations of the use of OH in food industry are mainly related to electric conductivity differences between liquid and solid phase sand the lack of appropriate temperature validating techniques for locating cold/hot spots. However, OH processing provides the production of shelf-stable products with high quality characteristics that can barely be achieved with traditional sterilization processes (Ohlsson 2002).

5.2. Dielectric Heating

5.2.1. Basic Principles

Dielectric heating, also known as high-frequency heating, radio frequency heating (RF) and microwave heating (MW) is a thermal process, which implies the interaction between an electromagnetic alternating field and the dipoles and ionic charges contained within a food product that enables the volumetric heating of the product. The principal mechanisms involved in dielectric heating are therefore dipole rotation and ionic polarization. In an alternating current electric field the polar molecules, such as water and ionic species, constantly realign themselves and are forced to reversing the electric field around the food product. The second major mechanism involved in dielectric heating is related with the polarization of ions as a result of the back and forth movement of the ionic molecules trying to align themselves with the oscillating electric field. The intern molecular friction due to molecular dipole rotation and migration of ionic species under the influence of the oscillating electromagnetic field generates heat inside the food by energy dissipation (Figure 2). Dielectric heating is also affected by the state of the food constituents and whether they are bound or free (Cheremisinoff 1996; Ahmed and Ramaswamy 2007).

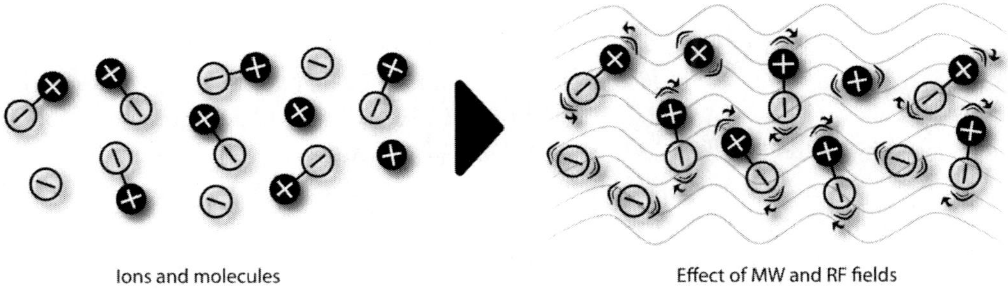

Figure 2. Schematic illustration of polar molecules in an electromagnetic field.

The heat power per unit volume in a particular location inside the food material can be given by:

$$G = 2\pi f \varepsilon_0 \varepsilon'' E^2 \quad \text{(Equation 2)}$$

The rate of heat generation (G) is directly proportional to the square of the electric field strength (E), to the microwave/radio frequency (f), to the permittivity of free space (ε_0) and to the dielectric loss factor representing the food materials ability to absorb the wave (ε").

5.2.2. Microwave Heating versus Radio Frequency Heating

Radio frequency and microwave are both capable of offering rapid and non-contact volumetric heating. They can be distinguished by the way that electric field is generated and depth of electromagnetic waves (Pereira and Vicente 2010).

In microwave heating, the electric field is generated by special oscillator tubes known as magnetrons or klystrons, which emit microwaves that are transferred by a wave guide into a metal chamber with the food product to be heated. It is a wave being propagated and reflected under the laws of optics (Figure 3a). In radio frequency heating the electric field is developed between electrodes by a triode valve. Although there are other configuration types, the most commonly used electrode configuration for heating thick materials is the parallel plat radio frequency heating system, in which the food product to be heated is placed between, but not touching, parallel electrodes (Figure 3b) (Giese and Darby 2000; Orsat and Raghavan 2005; Wang et al. 2005; Marra et al. 2009).

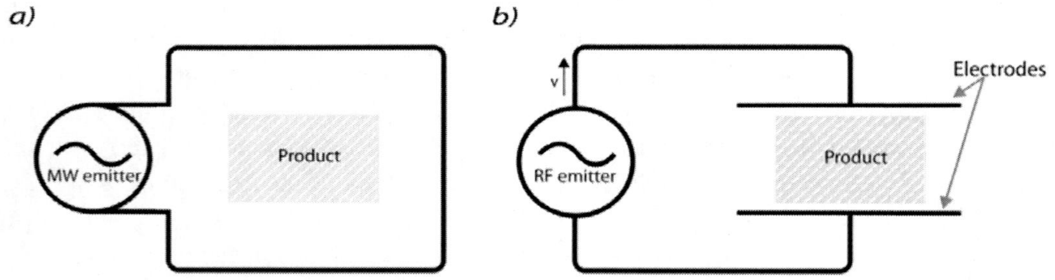

Figure 3. Schematic diagram of a) microwave and b) radio frequency heating systems.

Microwave heating is based on electromagnetic waves that are within a frequency band of 300 MHz to 3000 MHz. The microwave frequency bands used for industrial, scientific, and medical uses have been limited to a few narrow frequency bands to avoid interference with the radio frequency used for telecommunication purposes. Domestic microwave ovens generally operates at a frequency of 2450±50 MHz, while industrial systems typically operate at frequencies of 2450±50 MHz and 915±25 MHz with penetration depths ranging from 3 to 8 cm and 8 to 22 cm, respectively for each frequency, depending mainly on moisture content (Datta and Anantheswaran 2000).

On the other hand, the radio frequency heating uses electromagnetic energy between 1 and 300 MHz, although the main frequencies used for industrial heating lie in the range 10–50 MHz and only 13.56 ± 0.01, 27.12 ± 0.16 and 40.68 ± 0.02 MHz are permitted for industrial, scientific and medical applications (Tang et al. 2005; Marra et al. 2009). Radio frequency energy penetrates foods more deeply (penetration depth is generally greater than 1 m) than microwave energy, once wavelength at radio frequencies is up to 360 times greater than the wavelength of microwave treatments (Awuah et al. 2005).

5.2.3. Impact on Cellular Structures

The effect of microwave and radio frequency on chemical structures and microorganisms are mainly due to thermal effects. However, the mechanisms of microbial inactivation are not fully clarified and several theories have explained the process effects on cells, spores, or chemical compounds (Kozempel et al. 1998).

For microwave heating, the four theories that explain microorganisms inactivation are: *(i)* selective heating theory - the microorganisms are selectively heated due to microwave and reach higher temperatures than the surrounding medium, which lead to a rapid microbial inactivation; *(ii)* electroporation theory - electrical potential across cell membranes causes pores, resulting in the leakage of cellular materials; *(iii)* cell membrane rupture theory - the voltage applied across the cell membranes lead to the irrupture; and *(iv)* the magnetic field coupling theory - the internal cellular compounds are disrupted due to the coupling of electromagnetic energy with critical molecules such as protein or DNA (Kozempel et al. 1998).

In relation to radio frequency, the major impact on microbial inactivation is also due to heat generation on the substrate. Recently, several scientific publications have suggested non-heat associated mechanisms, such as improper protein folding, damages to the integrity of the membrane or DNA damages (Schlisselberg et al. 2013).

5.2.4. Principal Advantages

Microwave and radio frequency heating are good alternatives to conventional thermal treatments. One of the most import advantages is related with wave capacity to penetrate the food deeply and more uniformly than heat of conventional thermal methods. Consequently, the whole volume of food is processed more uniformly and more rapid, with significant reductions of processing time and energy (Ahmed and Ramaswamy 2007). Irregular heating may however occur, depending on food geometry and composition.

Other no less important advantage is related with food quality retention. Several quality parameters such as nutrients and vitamins contents, flavor, sensorial characteristics and color are satisfactory preserved due to the fast heat transfer process. Moreover, the surfaces of food products do not overheat, leading to minimal heat damages and avoiding surface browning. The heating efficiency of microwave is higher than the one of conventional methods and the processing is suitable for heat-sensitive, high-viscous, and multiphase fluids that cannot be processed with traditional thermal processes. The costs involved in microwave processing are also significant lower than the ones related to thermal processes (Fellows 2000; Ahmed and Ramaswamy 2007).

5.2.5. Food Industry Applications

Microwave processing has been successfully applied to (re) heating, baking, (pre) cooking, thawing and tempering, blanching, drying, pasteurizing and sterilizing foods, although its industrial application will always compete with thermal conventional processes (Fito et al. 2005; Sumnu and Sahin 2005; Regier et al. 2010). The heating of precooked foods is the main practical application of microwave ovens due to rapid, safe, and hygienic heating of the food product. The largest use of industrial microwave processing has been for tempering fish, meat and poultry. Some other industrial applications are: cooking of chicken and bacon (with advantage over oil frying), pastry products baking, fruits and vegetables blanching (with the advantage of avoiding the loss of nutrients and pigments, which are lixiviated in the blanching water), fruits and vegetables drying (with the advantages of reducing the process time, surface overheating and shrinkage) and pasteurization of pastry, prepared dishes, soft cheeses, milk and fruit juices.

Microwave sterilization is still only being applied at a laboratory scale. There are three main reasons that greatly limit the application of this process: *(i)* the measurement of temperatures at few locations do not guarantee the real temperature distribution of the product during microwave heating, lacking actual temperature profiles; *(ii)* metallic materials can not be used in packaging, because metallic surfaces reflect microwaves; and *(iii)* uncertainty about microbial inactivation (Fito et al. 2005; Sumnu and Sahin 2005; Ahmed and Ramaswamy 2007; Regier et al. 2010).

Radio frequency dielectric systems have been used mainly as heating and drying processes. One of the most important applications in the food industry is the post bake drying of cookies, biscuits, and crackers. However, radio frequency could also be particularly appropriate as a heat processing method for cured whole meat products (such as ham) and for drying food ingredients (such as herbs, spices and vegetables), snack foods, potato and pasta products. The radio frequency that allows a rapid and uniform heating are leading researchers for new applications, such as thawing of frozen foodstuffs, vegetables blanching, pasteurization and sterilization. However, before its industrial implementation, some questions such as dielectric arcing and thermal runaway heating need to be addressed (Tang et al. 2005).

6. Novel Non-thermal Food Processes

Food preservation methods that are not based on temperature action are called non-thermal processes. Processes such as high pressure, power ultrasounds, ozone treatments and UV-C radiation are innovative and reveal promising applications in food industry.

6.1. High Pressure Processing

6.1.1. Basic Principles

The high pressure processing (HPP) of foods involves the exposition of food products to pressures that generally can range from of 100 to 1000 MPa, from a millisecond pulse to over 20 min. During pressure treatment, the products temperature can be below 0ºC or above 100ºC depending of the food products requirements (Guan and Hoover 2005). The impact of HPP on microbial inactivation, chemical or enzymatic reactions, and structural and/or functional properties of food products are based on Pascal's isostatic principle and le Chatelier principle. Pascal's isostatic principle states that pressure applied to a sample (including biological products) is transmitted uniformly and instantaneously. Contrarily to traditional thermal methods, the pressurization process is independent of sample volume, product size and geometry. Le Chatelier principle specifies that any reaction or phase transition accompanied by a volume decrease is favored by high pressure, whereas reactions that involve a volume increase will be inhibited. In other words, the reactions accomplished by volume increase tend to be inhibited by pressure. Thus, pressure has a profound influence on the rates of the most part of the biochemical reactions that occur in foods, once they are often involved in volume changes. Moreover, pressure may affect the energy production of enzymatic reactions, inhibiting energy availability (Welti-Chanes et al. 2005; Venugopal 2006). High pressure can be generated by direct compression (requires dynamic pressure seals between piston and vessel surface), indirect compression (requires static pressure seals) or by heating the pressure medium, which is usually water combined with mineral or vegetable oil for lubrication with anticorrosive aims (Ohlsson 2002; Welti-Chanes et al. 2005).

6.1.2. Impact on Cellular Structures

HPP inactivates/denatures proteins and controls important enzymatic reactions responsible for food quality degradation (Eisenmenger and Reyes-de-Corcuera 2009). Pressure has a significant impact on water properties, such as thermal properties, superficial tension, density, viscosity, dipolar moment and dielectric constant. The compressibility of water is also affected by pressure, which influences microbial and enzymatic inactivation and promotes changes in the rheological properties of the food. When a high-moisture food is treated with HPP, the density of water and food increase. Small molecules or low-molecular weight biopolymers, those with covalent structures (such as peptides, lipids, and sugars) and the primary structures of macromolecules (such as proteins, nucleic acids, and polysaccharides) remain intact due to the negligible compressibility of the covalent bonds that do not break within the ranges of pressures normally used in food processing (Welti-Chanes et al. 2005). According to le Chatelier–Braun principle, the reactions that result in reduced volume will be promoted by pressure affecting the structure of large molecules such as proteins. Pressure can promote the disruption of the ionic bonds and hydrophobic interactions responsible for maintaining the secondary and tertiary structure of proteins (covalent and non-covalent interactions) during and upon release of pressure, thus triggering their denaturation. The rupture of ionic bonds, some hydrogen bonds and hydrophobic and electrostatic interactions change the secondary, tertiary or quaternary structure of large molecules and complex organized structures such as membranes (Rendueles et al. 2011). However, at moderate pressures (<101 MPa), the denaturation rate of proteins is reduced because the generation of hydrogen bonds, which are responsible for maintaining the helical structure of peptides, is accompanied by volume decrease that is favored by pressure (Welti-Chanes et al. 2005). The covalent bonds of food constituents are less affected than weaker bonds, possibly due to the low energy levels developed by pressure. Molecules such as vitamins, amino acids, flavor molecules or other low-molecular-weight compounds are hardly affected and, as a result, the organoleptic and nutritional properties slightly changed. Thus, there are no significant losses of sensory properties and nutrients, and food quality can be better preserved than in thermal methods (Rivalain et al. 2010; Mujica-Paz et al. 2011). In addition, HPP can inactivate the vegetative form of many microorganisms by interrupting cellular functions responsible for reproduction and survival. Cell death occurs due to a multiplicity of damages accumulated in different parts of the cell (cell permeability modifications and functional disruptions) that lead to irreversible leakage of intracellular compounds (Torres et al. 2010; Rendueles et al. 2011). The denaturation of key enzymes and the disruption of ribosomes are other mechanisms of action that may be responsible for microbial inactivation (Hogan et al. 2005). As a food preservation method, the effectiveness of HPP in destroying foodborne microorganisms depends on a number of intrinsic and extrinsic factors, such as the pressure applied, duration of compression, depressurization rate, nature of pressure treatment (cyclic or continuous), temperature of treatment, time of exposure, product pH, water activity and ionic strength (Venugopal 2006).

6.1.3. Principal Advantages

The major advantage of the HPP method is related with the extended shelf-life and improvements of food quality and safety aspects, due to microbial and enzymatic inactivation (Rendueles et al. 2011). Specific advantages of the process are: *(i)* homogeneity of the process, since it is an isostatic method and the pressure is uniformly applied around and throughout the food product; *(ii)* adiabatic temperature distribution throughout the product, *i.e.* the impact of HPP on food is instantaneous; *(iii)* high retention of fresh taste, sensory characteristics, texture, vitamins and other nutrients; *(iv)* efficient microbial inactivation; *(v)* the shelf-life of products is similar to thermal-pasteurized ones, however HPP allows maintaining natural food quality characteristics.

When compared to thermal methods, high pressure processing involves less energy requirements. A significant smaller amount of energy is required to compress a solid or liquid to 500 MPa than the amount of energy to heat up to 100°C.

A key advantage of HPP is its applicability to packaged foods. HPP cause minimum damage on food packaging due to isostatic effect making obsolete the efforts to prevent recontamination or the use of aseptic filling processes.

High pressure processing is also an environmentally friendly process with few effluents generated (Venugopal 2006; Pereira and Vicente 2010).

6.1.4. Combined Treatments

High pressure processing may be combined with other treatments to attain increased effectiveness and to reduce the deteriorative effects caused by excessive pressure.

Pressure assisted thermal sterilization (PATS) is a novel food sterilization methodology recently approved by U.S. Food and Drug Administration, which involves the simultaneous application of high pressures (500-800 MPa) and temperatures above 100ºC to a pre-heated foods (Ferreira et al. 2009; Nguyen et al. 2010; Ayvaz et al. 2012). This method uses the adiabatic heating effect caused by pressurization, to increase foods temperature instantaneously and uniformly. When foods are compressed, their temperature increases due to adiabatic heating effect (approximately 4-6ºC temperature increment per 100 MPa; Leadley et al. 2008). In addition, since the application of high pressure processing is nearly instantaneous and uniform, *i.e.* HPP effects on foods are independent of mass and nearly independent of time, the adiabatic heating has also those characteristics (Yaldagard et al. 2008; Rendueles et al. 2011). This means that a pre-heated food (~80-90ºC), when subjected to HPP, will have its temperature increased to sterilization values, nearly instantaneously and uniformly. After a holding time required to achieve sterilization, a decompression with the opposite effect occurs, *i.e.* adiabatic cooling, with the food temperature going back to the pre-heated values (~80-90ºC) (Rajan et al. 2006; Ramirez et al. 2009; Peleg et al. 2012).

This sterilization procedure leads too much shorter sterilization times and, consequently, a less impact on food quality is verified. However, microbial spores lethality is achieved and the processing costs are low (Rastogi et al. 2008). This opens the opportunity to sterilize a variety of heat sensitive products like egg products, mashed potatoes, and others, which are difficult or even impossible to sterilize by the traditional industrial retort process (Rastogi et al. 2007; Wimalaratne and Farid 2008).

Pressure-assisted freezing (PAF) consists in moderate pressure combined with temperatures under 0°C. PAF allows rapid freezing of foods without formation of large ice crystals, avoiding damage from formation of ice crystals and preventing microbial deterioration (Welti-Chanes et al. 2005).

Water is one important food constituent and, at certain pressures, it can exist in a non-frozen liquid state at temperatures below 0°C. For example, the freezing point of water exhibits a minimum temperature of -22°C at 207.5 MPa. Releasing pressure rapidly, a very high supercooling stage can be achieved as well as a very high ice nucleation rate, which leads to very small ice crystals (Hogan et al. 2005).

PAF has significant applications in food industry. For example beef and pork meat have been stored at temperatures ranging from 5°C to -20°C and pressures from 50–200 MPa (Welti-Chanes et al. 2005). Since 1992, several studies of PAF on fishery products, tofu and soy noodles, carrot, potato, eggplant, and cabbage have been developed, focusing mainly on food quality aspects. In general, results showed a significant increase of products quality (Le Bail et al. 2002; Venugopal 2006).

Pressure-assisted thawing (PAT) involves hawing at a lower temperature and under pressure. From a microbiological point of view, low temperature and high pressure cause a synergetic effect on microbial lethality. A faster thawing is also associated with less amount of drip formation when compared with conventional thawing. For example, based on heat transfer criterion, it was found that the optimum pressure for whiting thawing with temperatures ranging from -5 to -20°C was 210 MPa. Higher temperature and higher pressure caused partial melting during pressure loading, leading to temperature decrease of the product.

Some food products such as tuna meat, surimi blocks, beef, whole or sliced strawberries, chourot and pollock whiting fillets required shorter thawing times when PAT was used (Le Bail et al. 2002; Welti-Chanes et al. 2005).

6.1.5. Food Industry Applications

High pressure processing has become a commercially implemented technology worldwide, applied to a large variety of food products with consumers' great acceptability. High pressure processing can be applied to meat-based products (*e.g.* cooked and dry ham), seafood and fish, pre-cooked dishes, fruits and vegetables, cereals, milk and juices. The main applications of HPP are in the production of jams, fruit juices, soups, oysters and, more recently, processed meats such as hams (Hogan et al. 2005; Welti-Chanes et al. 2005). In the meat sector, HPP has been adopted for the post-packaging pasteurization of sliced meat products, including fermented meats, whole and sliced cured deli ham, dry cured ham, pre-cooked chicken strips, cold cuts and other delicatessen meat products. Other current applications are related with the fact that HPP induces the opening of shellfish shells. This technology has been applied to oysters, inactivating pathogenic microorganisms and retaining the sensory characteristics of fresh oysters for an extended shelf-life period. HPP has been applied successfully in other types of shellfish, such as mussels, *Nephrops*, and lobsters (Hayashi 1995; Hogan et al. 2005). This technology has also been used for soy protein hydrolysis, softening of meat, discoloration of hemoglobin or by-products of the meat industry, deodorizing of proteins, and solubilisation or modification fish proteins (Welti-Chanes et al. 2005).

The critical parameters of high pressure processing are: treatment pressure; come-up time to achieve target pressure; holding time at target pressure; decompression time; initial temperature of food products; process temperature; temperature distribution in the pressure vessel due to the adiabatic heating; characteristics of the food such as pH, composition and water content; type of contaminant microorganisms; packing material. The food size and shape do not affect the process, because pressure acts instantaneously and uniformly thought the product. However, there is a slight temperature increase with pressure increase, *i.e.* 3°C per 100 MPa, depending on food composition (Guan and Hoover 2005).

6.2. Power Ultrasounds

6.2.1. Basic Principles

Ultrasonic waves are similar to sound waves, but their frequencies are far too high for perception by human hearing. Normal human hearing detects sound frequencies between 0.016 kHz to 18.0 kHz and the power intensity of a quiet conversation is around 1 Wcm^{-2}. Although ultrasounds have the properties of sound waves, such as reflection, interference, adsorption, and scattering, and can be propagated through solids, liquids, and gases, they are classified as being of low and high energy (high and low frequency, respectively). Low intensity ultrasounds use very high frequencies between 2 to 20 MHz with low power levels from 100 $mWcm^{-2}$ to less than 1 Wcm^{-2}, while power ultrasounds usually use lower frequencies between 20 to 100 kHz and can produce much higher power levels, in the order of 10 to 1000 Wcm^{-2}. Energy intensities greater than 10 Wcm^{-2}, such as the ones used by power ultrasounds technology, break intermolecular bonds and generate cavitation effects, which are known to alter some physical properties as well as enhance or modify many chemical reactions due to the rapid formation and destruction of cavitation bubbles (Torley and Bhandari 1999; Leadley and Williams 2006).

In the liquid media, the best-known effect of ultrasound is cavitation. When ultrasound waves pass through a liquid, a continuous wave type motion is created. Longitudinal waves are propagated via a series of compressions and refractions waves, which are induced in the medium particles. If the wave has sufficient high amplitude, the rare faction cycle may exceed the attractive forces of the liquid molecules and gas bubbles or cavities are produced in the medium (cavitation process). These bubbles growth produce a phenomenon called rectified diffusion, *i.e.* during the expansion cycle the bubbles have a big surface area increased by the gas diffusion, causing the bubble expansion. That intake of gas during the expansion is not completely expelled during the compression cycle and, consequently, the bubbles grow. After some compression cycles, rapid condensation occurs, because the ultrasonic energy is not sufficient to retain the vapor phase in the bubble. The condensed bubble implodes violently creating shock waves and generating energy for chemical and mechanical effects (Figure 4) (Earnshaw et al. 1995; Ohlsson 2002; Piyasena et al. 2003). During the bubbles collapse, localized regions of very high temperature (up to 5500ºC) and pressures (up to 50 MPa) are created.

Ultrasonic waves have the capacity to produce different effects in different media and sometimes these effects seem contradictory. For example, power ultrasounds in liquid suspensions break aggregated particles leading to a very effective emulsification/ homogenization method but in gaseous suspensions tend to produce particle agglomeration (Mason et al. 2005; Leadley and Williams 2006). The success of the cavitation process depends on several factors such as the frequency of the vibration (increasing frequency increases the power required to produce cavitation), intensity of the vibration (related to amplitude), solvent viscosity, surface tension and vapor pressure, attenuation of the vibration by the medium (greater attenuation at higher frequency), the presence of gas bubbles as cavitation nuclei and ambient temperature and pressure (Torley and Bhandari 1999).

6.2.2. Impact on Cellular Structures

The destructive potential of ultrasounds was discovered in 1960's when it was observed that sound waves used in anti-submarine warfare killed fish. Since then, studies on how ultrasounds disrupt biological structures have been carried out, and numerous theories and mechanisms concerning ultrasound's harmful effects on cells have been proposed (Earnshaw et al. 1995; Piyasena et al. 2003). The antimicrobial effect of ultrasound technology is largely due to the localized, but extreme, pressures and temperatures produced during intracellular cavitation. This phenomenon promoted the disruption of cellular structures and functional components through cell walls damage (due to direct thermal effects from the localized heating), causing cell lysis, generation of free radicals that damage DNA, production of hydrogen peroxide and others chemical compounds, and microstreaming that causes thinning of cell membranes (Torley and Bhandari 1999; Butz and Tauscher 2002; Ross et al. 2003; Lee and Feng 2011).

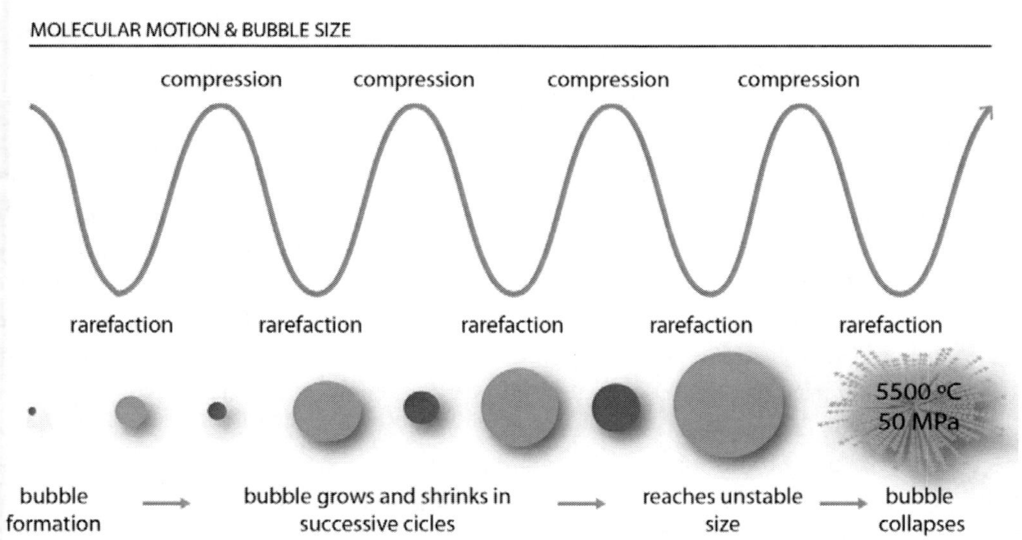

Figure 4. Ultrasonic cavitation and implosion.

The effectiveness of ultrasounds on microorganisms depends mainly on the type of bacteria and on the treatment time and temperature. Ultrasounds technology has been effective against a variety of microorganisms including bacteria, bacterial spores, yeasts, fungi, fungal spores, protozoa, and viruses. However, its action depends on the microorganism itself. The most resistant are spores and some viruses, which can be difficult to inactivate with ultrasounds. Enzymes can also be inactivated by ultrasounds due to a depolymerisation effect. The resistance of microorganisms and enzymes to ultrasounds, possibly due to the complexity of foods and their potential protective effect on microorganisms, can be very high and it may require several hours of ultrasonication to achieve desired results. Additionally, so prolonged treatments will produce extensive changes in the food products, being food quality severely affected. Those reasons make ultrasound a relatively inefficient process as preservation technique. However, its effectiveness in food preservation can be increased by combination with other treatments (Torley and Bhandari 1999; Mason et al. 2003).

6.2.3. Combined Treatments and Principal Advantages

Ultrasounds can be combined with heat and pressure, increasing the process efficiency in relation to enzymes and microbial inactivation. The use of ultrasounds combined with: *(i)* moderate heating is known as thermosonication; *(ii)* moderate pressure is known as manosonication and *(iii)* heat and pressure is known as manothermosonication. A limitation of thermosonication is related with its effectiveness decrease when temperature increases. The increase of vapor pressure and the decrease of liquid surface tension reduce the cavitation effect. The combination of ultrasounds and pressure overcomes this problem, because under pressure conditions, cavitation can occur at higher temperatures increasing the efficiency of microbial inactivation. Even at temperatures below the boiling point of the medium, the microbial inactivation is significantly affected by the application of pressure (Torley and Bhandari 1999).

Higher-power ultrasounds at low frequencies and applied alone is effective in vegetative cells inactivation. However, when associated with mild heat treatments, it increases the inactivation rate of several microorganisms, including spores. Additionally, the combination of ultrasounds to heat and pressure, has been reported to successfully inactivate heat resistant enzymes, cause cavitation and subsequent inactivate microorganisms at lower temperatures and/or in a shorter time than conventional thermal treatments at the same temperatures (Vercet et al. 1997; Piyasena et al. 2003; Knorr et al. 2004). Manothermosonication has been studied and applied at a laboratory scale. However, in a near future, this treatment may have the potential to be used as an alternative to pasteurisation or sterilization, mainly for liquid foods. In relation to solid foods or viscous materials, manothermosonication can be hardly used, because such products do not easily support cavitation and the energies required would be very high (Mason et al. 2003).

The combination of ultrasounds, temperature and pressure may cause additive, antagonist or synergistic effects on quality and safety characteristics of the products. However, in general these hurdle combinations are considered improved applications, when their results are compared to conventional thermal processes (Anonymous 2000; Butz and Tauscher 2002; Piyasena et al. 2003; López-Malo et al. 2005).

6.2.4. Food Industry Applications

One of the most important limiting factors of ultrasounds is the dispersion of void and pores that influence the monitoring and interpretation of data. However, the development of sensors, microprocessors and methods of signal analysis are facilitating the use of this non-thermal, non-destructive and environmental friendly technology. At an industrial level, it is necessary to study the frequency and power requirements for different products and to choose the best equipment and measurement tests (Mizrach 2008). The amplitude of ultrasonic waves, the contact time, the nature of microorganism/food, the volume of food to be processed, the food composition and the treatment temperature, are examples of some critical processing factors of concern. Nevertheless, the use of ultrasounds technology has increased in the food industry (Knorr et al. 2002).

Low intensity ultrasounds, at intensities lower than 1 W/m^2 and at a frequency higher than 100 kHz, are used for non-invasive detection (food control), providing information about physical-chemical properties of food materials and it is used in relaxation phenomena studies (Knorr et al. 2004). This treatment can be used to enhance the efficiency of whole cells without cell wall disruption, increasing the transport of nutrients or affecting membrane/seed permeability by microstreaming. It has also applications in ageing of fermented products and in the rapid induction of oxidation of alcoholic drinks for flavor development and early maturation (Leadley and Williams 2006).

High intensity ultrasounds, at intensities higher than 1 W/m^2 and at frequencies between 18 and 100 kHz, are regularly used for physical and chemical changes of food properties, such as extraction, emulsification, homogenization, separation, extrusion, cell disruption, degassing of liquid food, viscosity alteration, chemical reaction promotion (oxidation/reduction), enzyme and protein extraction, meat softening and modification of crystallization processes (McClements 1995; Villamiel and de Jong 2000; Knorr et al. 2004; Zheng and Sun 2006; Patist and Bates 2008). For example, the controlled crystallization of sugar solutions, hardening of fats, and the manufacture of chocolate and margarine can be improved by the application of power ultrasounds (Leadley and Williams 2006).

6.3. Aqueous or Gaseous Ozone

6.3.1. Basic Principles and Main Advantages

Ozone (O_3) is a powerful oxidizing agent formed in the stratosphere by sun light effect on oxygen molecules. Industrially, this molecule can be obtained by exposing oxygen molecules to corona discharges and to ultraviolet light; both procedures are similar to what happens in nature (Sharma 2005). The corona discharge is achieved by passing the feed gas between two closely spaced electrodes. When the feed gas (such as dry air, oxygen or a gaseous mixture) passes between the electrodes, the gas is partially ionized and subsequent atomic collisions produce ozone. The process of ozone generation by ultraviolet light is similar. The isolated oxygen atoms are formed by the photo-dissociation of oxygen caused by short wavelength UV radiation, around 240 nm. Thereafter, oxygen atoms and molecules interact to form ozone. Ozone can also be produced by water electrolysis, but the cost is considerably high when compared with the corona discharge (Graham 2000). Electrolysis is predominantly used by pharmaceutical and electronic industries.

Ozone gas has a blue tint at room temperature; at higher concentrations the color is not noticeable. It has a strong characteristic smell that can be described as that of "fresh air after a thunderstorm". Human olfactory senses are able to detect this odor, even when ozone is present at a very low concentration (< 0.01 ppm).

Ozone is thermally unstable and dissociates in a very short time to form oxygen or to react with other gases (Butz and Tauscher 2002; Güzel-Seydim et al. 2004a). Ozone is more soluble in water than oxygen with a half-life in distilled water of 20-30 minutes, before breaking down to oxygen. However, it is difficult to maintain ozone dissolved in water, especially if temperatures are high. Consequently, to uphold or increase its concentration in aqueous solution, it is necessary to have continuous ozone production and low temperatures (Graham 2000; Guzel-Seydim et al. 2004b). Ozone is considered a valuable disinfecting agent, due to its high oxidation power. However, its efficiency as oxidant depends greatly on the balance between ozone concentration and organic matter content (Sharma 2005). In the gaseous phase, ozone is more stable than in aqueous solutions.

6.3.2. Impact on Cellular Structures

Ozone is effective against some gram positive bacteria (such as *Listeria monocytogenes*, *Staphylococcus aureus*, *Bacillus cereus* and *Enterococcus feacalis*), against some gram negative bacteria (such as *Pseudomonas aeruginosa* and *Yersinia enterocolitica*), against some yeasts (*Candida albicans* and *Zygosaccharomyces bacilli*) as well as against some spores like *Aspergillusniger* (Restaino et al. 1995; Guzel-Seydim et al. 2004). Ozone's ability to inactivate such a large spectrum of microorganisms is due to its strong oxidation power. Figure 5 shows how ozone approaches/destroys the bacterial cell and induces subsequent cell lysis.

Ozone promotes progressive oxidation of vital cellular components; the cell surface is the initial ozone target. Ozone has the capacity to oxidize the double bonds of unsaturated lipids, resulting in cell disruption and subsequent leakage of cellular contents (Guzel-Seydim et al. 2004). Lipoprotein and lipopolysaccharide layers of gram negative bacteria are the first sites destroyed by ozone and consequently the cell permeability is strongly affected resulting in cell lysis. The nucleic acids are significantly damaged or even destroyed by ozone as is viral RNA. Ozone alters polypeptide chains in viral proteins coats (Kim et al. 1999).

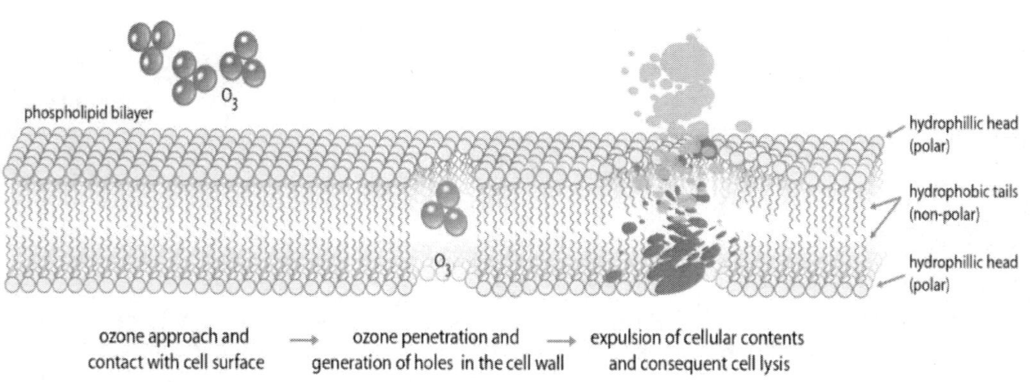

Figure 5. Effect of ozone on bacteria cell.

6.3.3. Food Industry Applications

Ozone was first used as a disinfectant of municipal waters, process waters, bottle drinking waters and swimming pools. Later, its application was extended to the treatment of wastewater, dairy and swine effluent, cooling towers, hospital water systems and equipment, aquariums and aquaculture, water theme parks and in public and home spas (Earth Tech Inc. 2005).

Some applications in the food industry include food preservation, surface hygiene, sanitation of food plant equipment, water disinfection and waste water reutilization, lowering biological oxygen demand (BOD) and chemical oxygen demand (COD) of food plant waste (Sheldon and Brown 1986; Schneider et al. 1991; Guzel-Seydim et al. 2004). Even at low concentrations and/or contact times, ozone can be effective in the inactivation of pure cultures of bacteria, molds, yeast, parasites and viruses in solution. However, the success of the food sanitation process is conditioned by the composition of food surface, type of microbial contaminant and the degree of attachment or association of microorganisms with food. In such situations, the food itself may compete with the microorganisms for ozone and may limit the accessibility of ozone to the microbial food contaminants (Kim et al. 1999; Sharma 2005). Ozone reacts with a large variety of organic compounds at very different rates, so there are many oxidation by-products that are formed during ozone processes. However, those by-products and residues are not potentially harmful and consequently, the ozonation process is considered safe.

Ozone, when dissolved in water, has been tested in post-harvest treatments of fruits and vegetables (Alexandre el at. 2011). However, it has greater efficacy when in the gaseous form, associated with refrigerated storage and modified atmosphere packaging. When used in water, ozone concentrations ranged from 0.03 to 20.0 ppm. When used in the gas form, the concentration reaches highly elevated doses such as 20000 ppm (Selma et al. 2008). Since ozone decomposition is rapid in water, its antimicrobial action may take place only at food surfaces (Aguayo et al. 2006).

6.4. Ultraviolet Radiation

6.4.1. Basic Principles and Main Advantages

The electromagnetic spectrum contains different forms of radiation that differ in penetrating power, frequency, and wavelength. Ultraviolet radiation (UV) covers a wide range of wavelengths between 100 and 400 nm of the electromagnetic spectrum. UV is categorized as UV-A (315 to 400 nm), UV-B (280 to 315 nm), UV-C (200 to 280 nm) and the vacuum UV range (100 to 200 nm). UV-A causes indirect damage to DNA, proteins and lipids through reactive oxygen intermediates; these are responsible for human skin changes that lead to tanning. UV-B and UV-C cause indirect and direct damage to DNA due to their strong power of absorption at wavelengths below 320 nm. UV-B can cause skin burning and eventually lead to skin cancer. UV-C is known to be germicidal; it inactivates bacteria, viruses, protozoa, yeasts, molds and algae (Bacon 2005).

Due to its strong germicidal effect, UV-C radiation has a great potential in food processing. Industrially, two types of UV lamps are of main interest: *(i)* the usual low-pressure mercury arc lamps with a monochromatic output at 254 nm (UV-C range), and *(ii)* the high intensity and medium pressure mercury arc lamps with a polychromatic output in all UV wavelengths (however concentrated at certain peaks within the UV-C region). Lamps that emit polychromatic light continuously have a medium light intensity. For a high light intensity, UV light pulses are often used (Giese and Darby 2000).

UV-C radiation can be considered a non-thermal food preservation technology, with the ability of minimizing the detrimental effects of traditional thermal methods and chemical sanitizers on quality and sensory attributes of food products. Some examples of positive aspects related to the application of UV-C radiation to foods are the absence of non toxic by-products, removal of some organic contaminants and low energetic requirements with minimal operating and maintenance costs. However, under particular circumstances, microorganisms may repair injury to UV-C light exposure and products' safety may be seriously compromised (Keyser et al. 2008). The lethal action of UV-C depends on the intensity of radiation, time of exposure, temperature, hydrogen ion concentration, number of organisms per unit area exposed and relative humidity. For relative humidity conditions higher than 50%, an increase in relative humidity leads to a decrease of the death rate of bacteria suspended in air. Apparently, fatty or waxy secretions on the cell surface of some mold species may protect them from UV-C radiation. Pigments may also provide some kind of protection; dark-pigmented spores are more resistant to UVlight than non-pigmented types.

6.4.2. Impact on Cellular Structures

Although the mechanisms of microbial inactivation due to UV radiation are not well understood, the germicidal effect of UV-C light has been demonstrated to be effective against some common foodborne pathogens such as *Campylobacter*, *Salmonella* and *Escherichia coli*.

A number of theories have been proposed for UV light action on cellular structures. Some researchers describe that the germicidal effect of radiation is mainly due to the photochemical reactions that are induced inside the microorganisms. The major microorganisms lesion is due to DNA absorption of the UV light (254 nm is the most lethal), taking place by cross-linking pyrimidine nucleoside bases (thymine and cytosine) in the same DNA strand. The formation of thymine dimmers in DNA and RNA will compromise cellular functions, since microorganisms can not perform the normal transcription and replication of the nucleic acids and, eventually, cell death occurs (Giese and Darby 2000; Wang et al. 2005; Altic et al. 2007; Sanz et al. 2007; Unluturk et al. 2008). UV radiation promotes the denaturation of proteins by modification or inactivation of functional enzymes and disruption of the functional components of nucleoproteins. Thus, the cells are overwhelmed by irreversible damage of multiple structures and functions, because they are unable to repair or reproduce critical proteins and repair DNA double helix structure (Green et al. 2003). However, some microorganisms may repair the damage caused by UV light exposure. Microorganisms can repair their DNA by light-independent (dark repair) and light-dependent (photo reactivation) mechanisms. Dark repair mechanisms occur through the nucleotide excision repair, post replication recombinational repair or by error-prone repair. The induction of those mechanisms depends on DNA damage. The photo reactivation repairs pyrimidine dimers in the DNA. This phenomenon is light dependent because uses a photolyase enzyme that can be activated in the presence of different wave lengths as UV-A and photosynthetic active radiation (400 to 700 nm) (Zenoff et al. 2006; Sanz et al. 2007).

Another theory defends that the cellular membrane damage, production of peroxyl radicals and concomitant exogenous photosensitization of microbial cells allowing absorption of peroxyl radicals may be attributed to the lethality of long-wave UV radiation (Bintsis et al. 2000). However, other authors refute the theory of production of oxygen radicals. In general, exposure to UV radiation required for microbial inactivation does not accelerate oxidative lipid processes (rancidity) that produces characteristics off-odors, off-colors, and textural changes; although organic peroxides may be involved (Doyle 1999).

Physical disintegration of bacterial cells or spores via intensive overheating of cells after absorption of pulsed UV light, is another theory (Green et al. 2003).

6.4.3. Food Industry Applications

The application of UV-C radiation may be divided into mainly in three categories: inhibition of microorganisms on surfaces, destruction of microorganisms in air, and liquids decontamination. Thus, industrially, it has been used in the disinfection of drinking water, waste water and air, as well as in the disinfection of surfaces and food containers (Marquenie et al. 2002; Wang et al. 2005; Altic et al. 2007). Bactericidal UV radiation have been applied for tenderizing or aging of meat, curing and wrapping of cheeses and prevention of surface mold growth on bakery products. Some juices, fruits and vegetables and fish can also be processed by UV light aiming at microbial load reduction. It has also been used in the ice cream industry and dairy processing plants (Guerrero-Beltrán and Barbosa-Cánovas 2004; López-Malo and Palou 2005; Unluturk et al. 2008). However, water treatment and surfaces sterilization are the fields where UV-C has been applied with a higher success due to the difficulty of light deep penetration beyond the opaque surface of several substances (Ohlsson 2002). This may limit the application of UV radiation to solid foods. Exposure to low doses of UV-C has been reported to reduce post-harvest decay of fruits and vegetables (Erkan et al. 2001; Marquenie et al. 2002; Allende and Artés 2003), which augurs potential application of UV-C radiation in the postharvest industry. However, low dosages may not be sufficient to inactivate some viruses and spores. Reported intensities for efficient UV treatments are around 400 J/m^2 (Anonymous 2000) and it is essential that the product surface is totally exposed to radiation. The critical parameters to be considered when designing a convenient UV radiation process for reduction microbial load and shelf life extension of foods are the type of food surface, food composition, intensity of radiation and exposure time (Anonymous 2000; Guerrero-Beltrán and Barbosa-Cánovas 2004).

CONCLUSION

Heat treatments are efficient in microbial inactivation, reducing products decay and attaining safety targets. Besides the proven effect of thermal treatments on microbial reduction, they have a significant impact on products' quality. Some quality parameters such as texture, color, aroma, flavor, taste and nutritive value can be severally affected by temperature. The application of low temperatures for long term food preservation are extensively used, however the size of ice crystals formed during freezing process affect mainly the texture of the products. The conventional washings treatments with chemical sanitizers have regulatory and mandatory restrictions. Sanitizer agents can be effective against a wide range of pathogens with minimal quality food loses, although in some circumstances, they may react with natural organic matter forming halogenated by-products, which can cause severe health problems.

The application of ohmic heating, microwave heating, radio frequency, high pressure processing, ultrasounds, aqueous or gaseous ozone and UV-C radiation are examples of mild or non-thermal thermal technologies that may have potential applications in the food industry. Some of these methods, such as high pressure processing and UV-C radiation, are already implemented in food industry, but only for some products and under specific conditions. These technologies should produce microbiologically safe products with increased shelf life, while reducing the process impact on quality characteristics. Moreover, mild or non-thermal treatments do not require a significantly consumption of energy, and therefore are more economic and environmental friendly technologies (Piyasena et al. 2003).

In ohmic or dielectric heating, the uniform heat generation results in a uniform temperature distribution and allows higher quality retentions of foods when compared to heat-based methods. However, the equipment investment costs may be significantly high.

In high pressure processing, the pressure is transmitted uniformly and instantaneously allowing the treatment of samples with different volumes, sizes, and geometry. However, although the process is economically viable, it is mainly suitable for high added value foods.

The ultrasound-based treatments are non-destructive and can be performed on-line, but they are very expensive to implement industrially.

Ozone treatments do not produce by-products and are more effective against microorganisms than the majority of sanitizers commonly used. However, as ozone is a strong oxidizing agent, its impact on humans is one of the most important parameters to be controlled. Industrially, it is essential to establish safe limits for people who are in contact with ozone during food processing and the use of masks with eyes protection should be recommended for the workers.

Radiation treatments, such UV-C light processes, do not have problems concerning by-products formation but, industrially, it is very difficult to measure the germicidal dose and, in some conditions, the microorganisms can repair their DNA and recover from injury.

Combination of such novel technologies with a mild thermal treatment, high pressure and traditional disinfectants solutions are promising. Combining two or more treatments, attaining target microbial inactivation without quality characteristics decay, can reduce the intensity of each technology and may conjointly improve safety and quality aspects of food products.

REFERENCES

Acuff, G. R. (2005). *Chemical decontamination strategies for meat: Improving the safety of fresh meat.* J. N. Sofos (eds). Woodhead Publishing Limited, Boca Raton, USA, 350-363.

Aguayo, E., Escalona, V. H. & Artés, F. (2006). Effect of cyclic exposure to ozone gas on physicochemical, sensorial and microbial quality of whole and sliced tomatoes. *Postharvest Biology and Technology,* 39(2): 169-177.

Ahmed, J. & Ramaswamy, H. S. (2007). *Microwave Pasteurization and Sterilization of Foods: Handbook of Food Preservation.* M. S. Rahman (eds). CRC Press, Taylor & Francis Group, Boca Raton, USA, 691-712.

Alexandre, E. M. C., Santos-Pedro, D. M., Brandão, T. R. S. & Silva, C. L. M. (2011). Influence of aqueous ozone, blanching and combined treatments on microbial load of red bell peppers, strawberries and watercress. *Journal of Food Engineering,* 105(2): 277-282.

Alexandre, E. M. C., Brandão, T. R. S. & Silva, C. L. M. (2013). Impact of non-thermal technologies and sanitizer solutions on microbial load reduction and quality factor retention of frozen red bell peppers. *Innovative Food Science & Emerging Technologies,* 17(0): 99-105.

Allende, A. & Artés, F. (2003). Combined ultraviolet-C and modified atmosphere packaging treatments for reducing microbial growth of fresh processed lettuce. *Lebensmittel-Wissenschaft und-Technologie,* 36(8): 779-786.

Alloway, B. J. (2004). Contamination of soils in domestic gardens and allotments: a brief overview. *Land Contamination & Reclamation,* 12(3): 179-187.

Altic, L. C., Rowe, M. T. & Grant, I. R. (2007). UV light inactivation of *Mycobacterium avium* subsp. paratuberculosis in milk as assessed by FASTPlaque TB Phage Assay and Culture. *Applied and Environmental Microbiology,* 73(11): 3728-3733.

Anonymous (2000). *Kinetics of microbial inactivation for alternative food processing technologies.* Center for food safety and applied nutrition. U. S. Food and Drug Administration.

Anonymous (2001). *Outbreaks associated with fresh produce: Incidence, growth and survival of pathogens in fresh and fresh-cut produce.* Analysis and evaluation of preventive control measures for the control and reduction/elimination of microbial hazards on fresh and fresh-cut produce. U. S. Food and Drug Administration.

Anonymous (2002). *Food safety and quality assurance issues: Improving the Safety and Quality of Fresh Fruits and Vegetables.* A Training Manual for Trainers (eds). University of Maryland: Food and Drug Administration. V1-V27.

Arino, A., Beltran, J., Herrera, A. & Roncales, P. (2005). *Fish: Encyclopedia of Human Nutrition.* 2nd Edition. B. Caballero, L. Allen and A. Prentice (eds). Elsevier Academic Press Ltd., UK, 247-256.

Awuah, G. B., Ramaswamy, H. S., Economides, A. & Mallikarjunan, K. (2005). Inactivation of *Escherichia coli* K-12 and *Listeria innocua* in milk using radio frequency (RF) heating. *Innovative Food Science & Emerging Technologies,* 6(4): 396-402.

Aymerich, T., Picouet, P. A. & Monfort, J. M. (2008). Decontamination technologies for meat products. *Meat Science,* 78(1–2): 114-129.

Ayvaz, H., Schirmer, S., Parulekar, Y., Balasubramaniam, V. M., Somerville, J. A. & Daryaei, H. (2012). Influence of selected packaging materials on some quality aspects of pressure-assisted thermally processed carrots during storage. *LWT - Food Science and Technology,* 46(2): 437-447.

Bachmann, J. & Earles, R. (2000). Postharvest Handling of Fruits and Vegetables, Horticulture Technical Note. *Appropriate Technology Transfer for Rural Areas,* 1-19.

Bacon, R. T. (2005). *Physical decontamination strategies for meat: Improving the safety of fresh meat.* J. N. Sofos (eds). Woodhead Publishing Limited, Boca Raton, USA, 318-349.

Behling, R. G., Eifert, J., M. C., E., Gurtler, J. B., Kornacki, J. L., Line, E., Radcliff, R., Ryser, E. T., Stawick, B. & Yan, Z. (2010). *Selected pathogens of concern to industrial food processors: infectious, toxigenic, toxico-infectious, selected emerging pathogenic bacteria: Principles of microbiological troubleshooting in the industrial food processing environment.* J. L. Kornacki (eds). Springer Science+Business Media, New York, USA, LLC 5-62.

Beuchat, L. R. (1998). *Surface decontamination of fruits and vegetables eaten raw: a review: Food safety issues.* Food Safety Unit (eds). World Health Organization,1-42.

Bintsis, T., Litopoulou-Tzanetaki, E., Davies, R. & Robinson, R. K. (2000). The antimicrobial effects of long-wave ultra-violet light and furocoumarins on some micro-organisms that occur in cheese brines. *Food Microbiology,* 17(6): 687-695.

Butz, P. & Tauscher, B. (2002). Emerging technologies: chemical aspects. *Food Research International,* 35(2-3): 279-284.

Cheremisinoff, N. P. (1996). *Dielectric heating: Microwave, radio frequency processes: Electrotechnology.* N. P. Cheremisinoff (eds). William Andrew Publishing, Westwood, 107-126.

Datta, A. K. & Anantheswaran, R. C. (2000). *Handbook of microwave technology for food applications.* A. K. Datta and R. C. Anantheswaran (eds). Marcel Dekker Inc, New York, USA.

Doyle, M. E. (1999). *Use of ultraviolet light to control Listeria in meat.* UV light Food Research Institute Letters.

Earnshaw, R. G., Appleyard, J. & Hurst, R. M. (1995). Understanding physical inactivation processes: combined preservation opportunities using heat, ultrasound and pressure. *International Journal of Food Microbiology,* 28(2): 197-219.

Earth Tech Inc. (2005). *Chlorine and alternative disinfectants.* Guidance Manual. Manitoba. http://www.gov.mb.ca/waterstewardship/odw/reginfo/approvals/odw_chlorine_and_alternative_disinfectants.pdf

Eisenmenger, M. J. & Reyes-de-Corcuera, J. I. (2009). High pressure enhancement of enzymes: A review. *Enzyme and Microbial Technology,* 45(5): 331-347.

Erkan, M., Wang, C. Y. & Krizek, D. T. (2001). UV-C irradiation reduces microbial populations and deterioration in Cucurbita pepo fruit tissue. *Environmental and Experimental Botany,* 45(1): 1-9.

FDA-CFSAN (2000). *Kinetics of microbial inactivation for alternative food processing technologies - Ohmic and inductive heating.* Food, Drug Administration-Center for Food Safety, Applied Nutrition.

Fellows, P. J. (2000). *Dielectric, ohmic and infrared heating: Food processing technology: Principles and practice.* P. J. Fellows (eds). Woodhead Publishing Limited and CRC Press LLC, Cambridge, England, 365-451.

Fennema, O. R., Powrie, W. D. & Marth, E. H. (1973). *Low temperature preservation of foods and living matter.* O. R. Fennema, W. D. Powrie and E. H. Marth(eds). Marcel Dekker, Inc, New York, USA.

Ferreira, E. H. R., Rosenthal, A., Calado, V., Saraiva, J. & S., M. (2009). *Byssochlamys nivea* inactivation in pineapple juice and nectar using high pressure cycles. *Journal of Food Engineering,* 95(4): 664-669.

Fito, P., Chiralt, A. & Martin, M. E. (2005). *Current state of microwave applications to food processing: Novel food processing technologies.* G. V. Barbosa-Cánovas, M. S. Tapia and M. P. Cano (eds). Marcel Dekker, New York, USA, 525-537.

Giese, N. & Darby, J. (2000). Sensitivity of microorganisms to different wavelengths of UV light: implications on modeling of medium pressure UV systems. *Water Research,* 34 (16): 4007-4013.

Gorny, J. (2006). *Microbial contamination of fresh fruits and vegetables: Microbiology of fruits and vegetables.* G. M. Sapers, J. R. Gorny and A. E. Yousef (eds). CRC Press, Taylor & Francis, Boca Raton, USA,3-32.

Goullieux, A. & Pain, J. P. (2005). *Ohmic heating: Emerging technologies for food processing.* D. W. Sun (eds). Elsevier Ltd, London, UK, 469-506.

Graham, D. M. (2000). *Ozone as an antimicrobial agent for the treatment, storage and processing of foods in gas and aqueous phases.* Direct food additive petition. D. M. Graham (eds). Palo Alto, California, USA.

Green, S., Basaran, N. & Swanson, B. G. (2003). *High-intensity light: Food preservation techniques.* P. Zeuthen and L. Bogh-Sorensen (eds). Woodhead Publishing Limited and CRC Press LLC, Cambridge, England, 284-302.

Guan, D. & Hoover, D. G. (2005). *Emerging decontamination techniques for meat: Improving the safety of fresh meat.* J. N. Sofos (eds). Woodhead Publishing Limited, Boca Raton, USA, 388-417.

Guerrero-Beltrán, J. A. & Barbosa-Cánovas, G. V. (2004). Advances and limitations on processing foods by UV light. *Food Science and Technology International,* 10(3): 137-147.

Güzel-Seydim, Z., Bever, P. I. & Greene, A. K. (2004a). Efficacy of ozone to reduce bacterial populations in the presence of food components. *Food Microbiology.* 21(4): 475-479.

Guzel-Seydim, Z. B., Greene, A. K. & Seydim, A. C. (2004b). Use of ozone in the food industry. *LWT - Food Science and Technology.* 37(4): 453-460.

Hartel, R. W. & Heldman, D. R. (1997). *Principles of Food Processing.* R. W. Hartel and D. R. Heldman (eds). Chapman & Hall, New York, USA.

Hayashi, R. (1995). *Advances in high pressure food processing technology in Japan: Food processing: Recent developments.* A. G. Gaonkar (eds). Elsevier Science B. V, Amsterdam, The Netherlands, 185-195.

Heldman, D. R. & Hartel, R. W. (1999). *Introduction: Principles of food processing.* D. R. Heldman and R. W. Hartel (eds). Aspen Publishers, Inc., Maryland, USA, 1-12.

Heldman, D. R. & Singh, R. P. (1981). *Thermodynamics of food freezing: Food process engineering.* D. R. Heldman and R. P. Singh (eds). Publishing Company, Inc., Westport, Connecticut, USA,158-215.

Hogan, E., Kelly, A. L. & Sun, D. W. (2005). *High pressure processing of foods: An overview: Emerging technologies for food processing.* D. W. Sun (eds). Elsevier Ltd., London, UK, 3-32.

Jakób, A., Bryjak, J., Wójtowicz, H., Illeová, V., Annus, J. & Polakovič, M. (2010). Inactivation kinetics of food enzymes during ohmic heating. *Food Chemistry,* 123(2): 369-376.

Keast, R. S. J. (2010). *Food quality perception: Processing effects on safety and quality of foods.* E. Ortega-Rivas (ed). CRC Press, Taylor & Francis Group, Boca Raton, USA, 67-86.

Kennedy, C. (2003). *Developments in freezing: Food preservation techniques.* P. Zeuthen and L. Bogh-Sorensen (eds). Woodhead Publishing Limited and CRC Press LLC, Cambridge, England, 228-240.

Keyser, M., Muller, I. A., Cilliers, F. P., Nel, W. & Gouws, P. A. (2008). Ultraviolet radiation as a non-thermal treatment for the inactivation of microorganisms in fruit juice. *Innovative Food Science & Emerging Technologies,* 9(3): 348-354.

Kim, J. G., Yousef, A. E. & Dave, S. (1999). Application of ozone for enhancing the microbiological safety and quality of foods: a review. *Journal of Food Protection,* 62: 1071-1087.

Knorr, D. (1995). *New methods of food preservation.* G. W. Gould (ed). Blakie Academic & Prof, London, UK.

Knorr, D., Ade-Omowaye, B. I. O. & Heinz, V. (2002). Nutritional improvement of plant foods by non-thermal processing. *Proceedings of the Nutrition Society,* 61(2): 311-318.

Knorr, D., Zenker, M., Heinz, V. & Lee, D.-U. (2004). Applications and potential of ultrasonics in food processing. *Trends in Food Science & Technology,* 15(5): 261-266.

Kozempel, M. F., Annous, B. A., Cook, R. D., Scullen, O. J. & Whiting, R. C. (1998). Inactivation of microorganisms with microwaves at reduced temperatures. *Journal of Food Protection,* 61(5): 582-585.

Lawrie, R. A. (1985). *Meat science - Fourth edition.* R. A. Lawrie(eds). Oxford, New York, Toronto, Sydney, Paris, Frankfurt: Pergamon Press.

Leadley, C., Tucker, G. & Fryer, P. (2008). A comparative study of high pressure sterilisation and conventional thermal sterilisation: Quality effects in green beans. *Innovative Food Science & Emerging Technologies,* 9(1): 70-79.

Leadley, C. E. & Williams, A. (2006). *Pulsed electric field processing, power ultrasound and other emerging technologies: Food processing handbook.* J. G. Brennan (ed). Wiley-VCH Verlag GmbH & Co. KGaA, Weinheim, Germany, 201-235.

le Bail, A., Chevalier, D., Mussa, D. M. & Ghoul, M. (2002). High pressure freezing and thawing of foods: a review. *International Journal of Refrigeration,* 25(5): 504-513.

Lee, H. & Feng, H. (2011). *Effect of power ultrasound on food quality: Ultrasound technologies for food and bioprocessing.* H. Feng, G. V. Barbosa-Cánovas and J. Weiss (eds). Springer Science+Business Media, LLC, New York, USA, 559-562.

Lima, M. (1999). *Food preservation aspects of ohmic heating: Handbook of food preservation.* M. S. Rahman (ed). Marcel Dekker, New York, USA, 741-750.

López-Malo, A. & Palou, E. (2005). *Ultraviolet light and food preservation: Novel food processing technologies.* G. V. Barbosa-Cánovas, M. S. Tapia and M. P. Cano (eds). Marcel Dekker, New York, USA, 405-421.

López-Malo, A., Palou, E., Jiménez-Fernández, M., Alzamora, S. M. & Guerrero, S. (2005). Multifactorial fungal inactivation combining thermosonication and antimicrobials. *Journal of Food Engineering,* 67(1-2): 87-93.

Marquenie, D., Michiels, C. W., Geeraerd, A. H., Schenk, A., Soontjen, C., van Impe, J. F. & N icolăi, B. M. (2002). Using survival analysis to investigate the effect of UVC and heat treatment on storage rot of strawberry and sweet cherry. *International Journal of Food Microbiology,* 73(2-3): 187-196.

Marra, F., Zhang, L. & Lyng, J. G. (2009). Radio frequency treatment of foods: Review of recent advances. *Journal of Food Engineering,* 91(4): 497-508.

Martins, R. C. & Silva, C. L. M. (2001). Modelling colour and chlorophyll loss of frozen green beans (*Phaseolus vulgaris,* L.). *International Journal of Refrigeration,* 25(7): 966-974.

Mason, T. J., Paniwnyk, L. & Chemat, F. (2003). *Ultrasound as a preservation technology: Food preservation techniques.* P. Zeuthen and L. Bogh-Sorensen (eds). Woodhead Publishing Limited and CRC Press LLC, Cambridge, England, 303-337.

Mason, T. J., Riera, E., Vercet, A. & Lopez-Buesa, P. (2005). *Application of ultrasound: Emerging technologies for food processing.* D. W. Sun (ed). Elsevier Ltd., London, UK, 323-351.

McClements, D. J. (1995). Advances in the application of ultrasound in food analysis and processing. *Trends in Food Science & Technology,* 6(9): 293-299.

McKenna, B. M., Lyng, J., Brunton, N. & Shirsat, N. (2006). Advances in radio frequency and ohmic heating of meats. *Journal of Food Engineering,* 77(2): 215-229.

Mizrach, A. (2008). Ultrasonic technology for quality evaluation of fresh fruit and vegetables in pre- and postharvest processes. *Postharvest Biology and Technology,* 48(3): 315-330.

Mujica-Paz, H., Valdez-Fragoso, A., Samson, C.T., Welti-Chanes, J. & Torres, J.A. (2011). High-Pressure Processing Technologies for the Pasteurization and Sterilization of Foods. *Food and Bioprocess Technology,* 4(6): 969-985.

Nguyen, L. T., Tay, A., Balasubramaniam, V. M., Legan, J. D., Turek, E. J. & Gupta, R. (2010). Evaluating the impact of thermal and pressure treatment in preserving textural quality of selected foods. *LWT - Food Science and Technology,* 43(3): 525-534.

North, M. F. & Lovatt, S. J. (2012). *Chilling and freezing meat: Handbook of meat and meat processing.* Second edition. Y. H. Hui (ed). CRC Press, Taylor & Francis Group, Boca Raton, USA, 357-380.

Ohlsson, T. (2002). *Minimal processing of foods with non-thermal methods: Minimal processing technologies in the food industry.* T. O. A. N. Bengtsson (ed). Woodhead Publishing Limited, Cambridge, UK,4-60.

Orsat, V. &Raghavan, G. S. V. (2005). *Radio-frequency processing: Emerging technologies for food processing.* D. W. Sun (ed). Elsevier Ltd., London, UK, 445-468.

Patist, A. & Bates, D. (2008). Ultrasonic innovations in the food industry: From the laboratory to commercial production. *Innovative Food Science & Emerging Technologies,* 9(2): 147-154.

Peleg, M., Corradini, M. G. & Normand, M. D. (2012). On quantifying nonthermal effects on the lethality of pressure-assisted heat preservation processes. *Journal of Food Science,* 71(1): R47-R56.

Pereira, R. N. & Vicente, A. A. (2010). Environmental impact of novel thermal and non-thermal technologies in food processing. *Food Research International,* 43(7): 1936-1943.

Piyasena, P., Mohareb, E. & McKellar, R. C. (2003). Inactivation of microbes using ultrasound: a review. *International Journal of Food Microbiology,* 87(3): 207-216.

Potter, N. N. & Hotchkiss, J. H. (1998). *Food Science - Fifth edition.* N. N. Potter and J. H. Hotchkiss(eds). Aspen Publishers, Inc., Gaithersburg, Maryland, USA.

Qi, B., Zang, Q., Barbosa-Cánovas, G. V. & Pedrow, P. D. (1995). Transactions of the ASAE. *Food and Process Engineering,* 557-565.

Rajan, S., Pandrangi, S., Balasubramaniam, V. M. & Yousef, A. E. (2006). Inactivation of *Bacillus stearothermophilus* spores in egg patties by pressure-assisted thermal processing. *LWT - Food Science and Technology,* 39(8): 844-851.

Ramirez, R., Saraiva, J., Lamela, P. C. & Torres, J. A. (2009). Reaction kinetics analysis of chemical changes in pressure-assisted thermal processing. *Food Engineering Reviews,* 1(1): 16-30.

Rastogi, N. K., Nguyen, L. T. & Balasubramaniam, V. M. (2008). Effect of pretreatments on carrot texture after thermal and pressure-assisted thermal processing. *Journal of Food Engineering,* 88(4): 541-547.

Rastogi, N. K., Raghavarao, K. S. M. S., Balasubramaniam, V. M., Niranjan, K. & Knorr, D. (2007). Opportunities and challenges in high pressure processing of foods. *Critical Reviews in Food Science and Nutrition,* 47(1): 69-112.

Regier, M., Rother, M. & Schuchmann, H. P. (2010). *Alternative heating technologies: Processing effects on safety and quality of foods.* E. Ortega-Rivas(ed). CRC Press, Taylor & Francis Group, Boca Raton, USA, 187-230.

Rendueles, E., Omer, M. K., Alvseike, O., Alonso-Calleja, C., Capita, R. & Prieto, M. (2011). Microbiological food safety assessment of high hydrostatic pressure processing: A review. *LWT - Food Science and Technology,* 44(5): 1251-1260.

Restaino, L., Frampton, E. W., Hemphill, J. B. & Palnikar, P. (1995). Efficacy of ozonated water against various food-related microorganisms. *Applied and Environmental Microbiology,* 61(9): 3471-3475.

Rico, D., Martín-Diana, A. B., Barat, J. M. & Barry-Ryan, C. (2007). Extending and measuring the quality of fresh-cut fruit and vegetables: a review. *Trends in Food Science & Technology,* 18(7): 373-386.

Rivalain, N., Roquain, J. & Demazeau, G. (2010). Development of high hydrostatic pressure in biosciences: Pressure effect on biological structures and potential applications in Biotechnologies. *Biotechnology Advances,* 28(6): 659-672.

Ross, A. I. V., Griffiths, M. W., Mittal, G. S. & Deeth, H. C. (2003). Combining nonthermal technologies to control foodborne microorganisms. *International Journal of Food Microbiology,* 89(2-3): 125-138.

Sagoo, S. K., Little, C. L., Gillespie, I. A. & Mitchell, R. T. (2003). Microbiological study of ready-to-eat salad vegetables from retail establishments uncovers a national outbreak of Salmonellosis. *Journal of Food Protection,* 66: 403-409.

Sanz, E. N., Dávila, I. S., Balao, J. A. A. & Alonso, J. M. Q. (2007). Modelling of reactivation after UV disinfection: Effect of UV-C dose on subsequent photoreactivation and dark repair. *Water Research,* 41(14): 3141-3151.

Sapers, G. M. & Simmons, G. F. (1998). Hydrogen peroxide disinfection of minimally processed fruits and vegetables. *Food Technology,* 52(2): 48-52.

Sastry, S. K. (2005). *Advances in ohmic heating and moderate electric field (MEF) processing: Novel food processing technologies.* G. V. Barbosa-Cánovas, M. S. Tapia and M. P. Cano(eds). Marcel Dekker, New York, USA, 491-500.

Schlisselberg, D. B., Kler, E., Kalily, E., Kisluk, G., Karniel, O. & Yaron, S. (2013). Inactivation of foodborne pathogens in ground beef by cooking with highly controlled radio frequency energy. *International Journal of Food Microbiology,* 160(3): 219-226.

Schneider, K. R., Steslow, F. S., Sierra, F. S., Rodrick, G. E. & Noss, C. I. (1991). Ozone depuration of Vibrio vulnificus from the southern quahog clam, Mercenaria campechiensis. *Journal of Invertebrate Pathology,* 57(2): 184-190.

Selma, M. V., Ibáñez, A. M., Cantwell, M. & Suslow, T. (2008). Reduction by gaseous ozone of *Salmonella* and microbial flora associated with fresh-cut cantaloupe. *Food Microbiology,* 25(4): 558-565.

Sharma, R. (2005). *Ozone descontamination of fresh fruits and vegetables: Improving the safety of fresh fruit and vegetables.* W. Jongen (ed). Woodhead Publishing Limited, Cambridge, UK, 373-386.

Sheldon, B. W. & Brown, A. L. (1986). Efficacy of ozone as a disinfectant for poultry carcasses and chill water. *Journal of Food Science,* 51(2): 305-309.

Singh, R. P. & Heldman, D. R. (2009). *Food freezing: Introduction to food engineering.* R. P. Singh and D. R. Heldman (eds). Academic Press, Elsevier, California, USA, 501-541.

Skip, R. & Seward, A. (2003). *Definition of food safety: Food safety handbook.* R. H. Schmidt and G. E. Rodrick (eds). John Wiley & Sons, Inc., New Jersey, USA, 3-10.

Stopforth, J. D. & Sofos, J. N. (2005). *Carcass chilling: Improving the safety of fresh meat.* J. N. Sofos (ed). Woodhead Publishing Limited, Boca Raton, USA, 364-387.

Sumnu, G. & Sahin, S. (2005). *Recent developments in microwave heating: Emerging technologies for food processing.* D. W. Sun (ed). Elsevier Ltd., London, UK, 417-444.

Tang, J., Wang, Y. & Chan, T. V. C. T. (2005). *Radio-frequency heating in food processing: Novel food processing technologies.* G. V. Barbosa-Cánovas, M. S. Tapia and M. P. Cano (eds). Marcel Dekker, New York, USA, 500-524.

Torley, P. J. & Bhandari, B. R. (1999). *Ultrasound in food processing and preservation: Handbook of food preservation.* M. Shafiur Rahman (ed). Marcel Dekker, New York, USA, 713-739.

Torres, J. A., Sanz, P. D., Otero, L., Lamela, M. C. P. & Saldaña, M. D. A. (2010). *Improving food safety and quality by high pressure processing: Processing effects on safety and quality of foods.* E. Ortega-Rivas (ed). CRC Press, Taylor & Francis Group, Boca Raton, USA, 379-414.

Tucker, G. S. (2011). *Food biodeterioration and methods of preservation: Food and beverage packaging technology.* R. Coles and M. J. Kirwan (eds). Wiley-Blackwell, West Sussex, UK.

Unluturk, S., AtIlgan, M. R., Handan Baysal, A. & TarI, C. (2008). Use of UV-C radiation as a non-thermal process for liquid egg products (LEP). *Journal of Food Engineering,* 85(4): 561-568.

Varnam, A. H. & Sutherland, J. P. (1995). *Meat and meat products: Technology, chemistry and microbiology.* A. H. Varnam and J. P. Sutherland (eds). Chapman & Hall, London, UK.

Venugopal, V. (2006). *High pressure processing: Seafood processing - Adding value* through quick freezing, retortable packaging, and cook-chilling. V. Venugopal (ed). Taylor & Francis Group, LLC, Boca Raton, USA, 319-340.

Vercet, A., Lopez, P. & Burgos, J. (1997). Inactivation of heat-resistant lipase and protease from *Pseudomonas fluorescens* by nanothermosonication. *Journal of Dairy Science,* 80(1): 29-36.

Villamiel, M. & de Jong, P. (2000). Inactivation of *Pseudomonas fluorescens* and *Streptococcus thermophilus* in Trypticase® Soy Broth and total bacteria in milk by continuous-flow ultrasonic treatment and conventional heating. *Journal of Food Engineering,* 45(3): 171-179.

Waite, J. G. & Yousef, A. E. (2010). *Overview of food safety: Processing effects on safety and quality of foods.* E. Ortega-Rivas (ed). CRC Press, Taylor & Francis Group, Boca Raton, USA, 11-66.

Wang, T., MacGregor, S. J., Anderson, J. G. & Woolsey, G. A. (2005). Pulsed ultra-violet inactivation spectrum of *Escherichia coli. Water Research,* 39(13): 2921-2925.

Warriner, K. (2005). *Pathogens in vegetables: Improving the safety of fresh fruit and vegetables.* W. Jongen (eds). Woodhead Publishing Limited, Cambridge, UK, 3-43.

Welti-Chanes, J., López-Malo, A., Palou, E., Bermúdez, D., Guerrero-Beltrán, J. A. & Barbosa-Cánovas, G. V. (2005). *Fundamentals and applications of high pressure processing to foods: Novel food processing technologies.* G. V. Barbosa-Cánovas, M. S. Tapia and M. P. Cano (eds). Marcel Dekker, New York, USA,157-152.

Wimalaratne, S. K. & Farid, M. M. (2008). Pressure assisted thermal sterilization. *Food and Bioproducts Processing,* 86(4): 312-316.

Yaldagard, M., Mortazavi, S. A. & Tabatabaie, F. (2008). The principles of ultra high pressure technology and its application in food processing/preservation: A review of microbiological and quality aspects. *African Journal of Biotechnology,* 7(16): 2739-2767.

Yildiz-Turp, G., Sengun, I. Y., Kendirci, P. & Icier, F. (2013). Effect of ohmic treatment on quality characteristic of meat: A review. *Meat Science,* 93(3): 441-448.

Zell, M., Lyng, J. G., Cronin, D. A. & Morgan, D. J. (2009). Ohmic cooking of whole beef muscle – Optimisation of meat preparation. *Meat Science,* 81(4): 693-698.

Zell, M., Lyng, J. G., Cronin, D. A. & Morgan, D. J. (2010). Ohmic cooking of whole turkey meat – Effect of rapid ohmic heating on selected product parameters. *Food Chemistry,* 120(3): 724-729.

Zenoff, V. F., Sineriz, F. & Farías, M. E. (2006). Diverse responses to UV-B radiation and repair mechanisms of bacteria isolated from high-altitude aquatic environments. *Applied and Environmental Microbiology,* 72(12): 7857-7863.

Zhao, Y. (2005). *Pathogens in fruit: Improving the safety of fresh fruit and vegetables.* W. Jongen (eds). Woodhead Publishing Limited, Cambridge, UK, 44-88.

Zheng, L. & Sun, D.-W. (2006). Innovative applications of power ultrasound during food freezing processes: a review. *Trends in Food Science & Technology,* 17(1): 16-23.

In: Processed Foods
Editors: Chloe M. Gagne and Daniel B. Jones
ISBN: 978-1-62808-688-1

Chapter 2

PERSISTENCE OF TRANSGENIC DNA IN PROCESSED FOODS OF ANIMAL ORIGIN: A POTENTIAL BIOLOGICAL RISK

G. Vinci**, *A. Tieri and R. Preti
Laboratory of Commodity Sciences, Department of Management,
Sapienza University of Rome, Italy

ABSTRACT

The consumer awareness on the use of genetically modified organisms (GMOs) in food production is growing rapidly, as well as the demand for total transparency.

The authorization of GMOs raw materials for further processing into animal feed or food is subject to strict European standards, and the labeling of their presence just above 0.9% is mandatory.

But there is no obligation to inform the consumer about their use in the production process. It follows that the consumer may be informed of the presence of GMOs in food only when it exceeds the threshold value of 0.9%, while being completely unaware if animal products were obtained by feeding the animals with even 100% GMOs.

Processed foods have gained popularity in recent years, many of the most produced being of animal origin, such as dairy products, meat and fish.

One of the most studied GM food safety concern is the fate of ingested recombinant DNA throughout the food chain, from GM crops through the gastrointestinal tract (GIT) of livestock, then to animal tissues and offspring and subsequently to humans. The main biological risk is the occurrence of horizontal transgenic transfer (HGT) to bacteria in the host GIT or animal cells, with many, almost unexplored implications.

The goal of this review is to organize and summarize the significant amount of research regarding the fate of dietary DNA in livestock and its persistence in animal origin processed food. The main technique of investigation is Polymerase Chain Reaction (PCR). Many results of recent publications agree that most of the DNA is degraded in the GIT by endogenous restriction nucleases, nevertheless some DNA fragments have been found in animal tissues and in products such as milk and meat.

* e-mail: giuliana.vinci@uniroma1.it.

Many discordant results have been reported about the presence of transgenic DNA, due to methodological limitations. Therefore additional research is needed in the field of both genetic traceability studies and DNA quantification methods development.

1. Introduction

GMOs are either microorganisms, plant or animal, in which their genetic compositions have been altered by a gene modification technique to gain new characteristics, such as herbicide tolerance, insect resistance, an enhanced shelf life or modified nutritional composition (Greiner & Konietzny, 2008; Lin, Chiueh, & Shih, 2000).

Crop plants are most widely used for this purpose and are more commercialized than other organisms. Soybean, maize, cotton and canola are the most cultivated GM plants, occupying 47%, 32%, 15% and 5% of the global GM crop area, respectively Soy and maize have a special importance because they represent the staple constituents of many foods. The planting of GM crops worldwide has constantly increased (94-fold) since that first crop and has reached 160 million hectares in global are according to the final reports of 2011 The major producers of GM crops are the United States, Brazil, Argentina, India, Canada and China (James, 2011).

In the EU the use of GMOs is subject to specific authorizations, as a guarantee for the environment and safeguard of human and animal health (European Union Commission, 2003). While permissions for cultivation are very strict, as already mentioned, position is different for consumption, in which case the list of authorized GMOs is far wider, not preventing imports of transgenic products in the EU from third countries. The authorization of GMOs raw materials for further processing into animal feed or food, can bypass the EU Commission through the EFSA to be more rapidly approved.

Even if the current European standards allow the use of authorized GMOs only, it should be stressed that is mandatory to label their presence just above 0.9%, while there is no obligation to inform the consumer about their use in the production process. It follows that the consumer may be informed of the presence of GMOs in food only when it exceeds the threshold value of 0.9%, while being completely unaware if animal products were obtained by feeding the animals with even 100% GMOs. However, the presence of transgenes in animal products could affect commodity sales. As different surveys carried out in 2010 report (Eurobarometer, 2010), more than 60% of European consumers are against GMOs' use and they are not willing to buy GMOs or food containing GMOs. Therefore from this point of view the use of GMOs has been diverted to the field of animal feed.

The consumers clearly opt for products GMOs free and foodstuffs in which GMOs are not involved at any stage of production, therefore they want and they need to be informed so that they can freely choose what to buy and what to eat can make their conscious choice. For the reasons stated above, a general trend of devaluation is a real risk, especially for high quality food products affected by the presence and/or use of GMOs. Such a depreciation penalizes both the farmers, because their profit margins will decline further, and the consumers, who will move towards more expensive certified GMO-free products as soon as their income will increase (Eurostat, 2011).

Interest in the consumption of transgenic protein and DNA has prompted investigations of their fate within the gastrointestinal tract of livestock and the potential to which transgenes or their products may be incorporated into tissues. Transgenic protein has not been detected in any animal tissues or products. Fragments of DNA from endogenous, high-copy number chloroplast genes from plants have been detected in poultry, pig and ruminant tissues. Low-copy endogenous and transgenic DNA in animal tissues have been detected but to a lesser extent than high-copy genes. Current research suggests that the passage of dietary DNA fragments across the intestinal wall is a natural physiological event, dependent on their amount in the feed. (Mazza et al. 2005; Rizzi et al, 2012)

According to the European Commission safety evaluation required to approval of importation or cultivation of a novel crop, GM plants has to be tested for substantial equivalence as risk assessment for humans, animals, and environment.

The aim of such a test is to determine whether a transgenic plant is substantially equivalent to its conventional counterpart at a chemical and dietary level, but the protocol for establishing substantial equivalence does not complete a nutritional safety assessment. For livestock nutrition, the measured parameters include crude protein, total lipids, fatty acids, fiber, starch, amino acids, metal ashes and sugars. These factors not only affect animal health and performance, but can also alter the composition and quality of animal products provided to the consumer. To date, no studies have demonstrated adverse effects in any of the currently registered GM plants or products derived from them used as feeds.

Food and Agriculture Organization (FAO) and World Health Organization (WHO) have both stated that there is no direct health risk to consumers ingesting GMO since the DNA from all organisms is structurally similar for the genetic code universality principle (FAO/WHO, 2000).

For the stable transfer of plant DNA into a microbial or mammalian cell to occur: (1) the relevant gene has to be released as linear fragment; (2) the gene has to avert nucleases cleavage both in the plant and in the gastrointestinal tract; (3) the gene has to compete with dietary DNA for uptake; (4) the recipient bacteria or mammalian cells have to transform the exogenous DNA; (4) the gene has to survive the hosts' restriction enzymes activity; (5) the gene has to be inserted into the host DNA (by repair or recombination); (6) the eventually inserted gene will be significantly diluted in microbial DNA and would have to compete with that DNA for uptake into a cell, especially in animals that harbour substantial digestive microbial populations upstream of gastric digestion in their GIT, such as ruminants and poultry (WHO, 2002).

Moreover, the gene has to survive any feed processing events prior to intake by livestock as well. In addition, once ingested, DNA is reduced to smaller fragments, nucleotides, and nucleosides as a result of mastication, acid hydrolysis, and enzymatic cleavage during digestion (Mc Allan, 1982); furthermore, it has to compete with other plant DNA (Beever and Kemp, 2000).

However, knowledge about the fate of ingested feed DNA fragments is still limited.

In processed food of animal origin the problem of GM DNA presence has two sources: the possible persistence of the recombinant DNA along the food chain, from the GM feed to the animal origin food, but also in the addition of GM proteins (common is from soybean) to processed meat products (hot-dog sausages, mortadella, cooked ham and others) that has significantly increased in recent years because of the interest in the functional and nutritional properties of these proteins (Belloque et al., 2002; Brod and Arisi, 2007).

2. RECOMBINANT DNA INTAKE BY LIVESTOCK

It has to be stressed that DNA ingested depends on the concentration of transgene in feed as well as feed intake. According to the main investigations, the quantity of DNA in most crops is less than 0.2 g/kg Dry Matter (DM) (Beever and Kemp, 2000). As a percentage of total DNA, the transgene concentration is unlikely to change for an individual event because nuclear gene insertions are generally static. However, as a percentage of DM intake, the concentration of transgenic DNA can vary in response to environmental conditions that change the overall biomass of the plant, such as lignifications with aging. Different parts of a plant will have variable transgenic DNA content as well, which is dependent on the quantity of cells per gram of DM. As an example, Beever and Phipps (2001) estimated that a dairy cow consuming 24 kg DM/day of a diet containing 400 g transgenic maize silage and 200 maize grain/kg, would have an intake of 57 g/day of total plant DNA. Of that, 54 g would be recombinant DNA and account for only 0.00000094 % of the total DNA intake. The actual total transgenic DNA intake may actually be lower, considering that ensiling GM plants quickly leads to degradation of large plant DNA fragments and that other processes, such as heat treatment, also lower the intake of intact transgenes. After ensiling, the quantity of each gene decreased to less than 0.03% of the starting amount. (Hupfer et al., 1999).

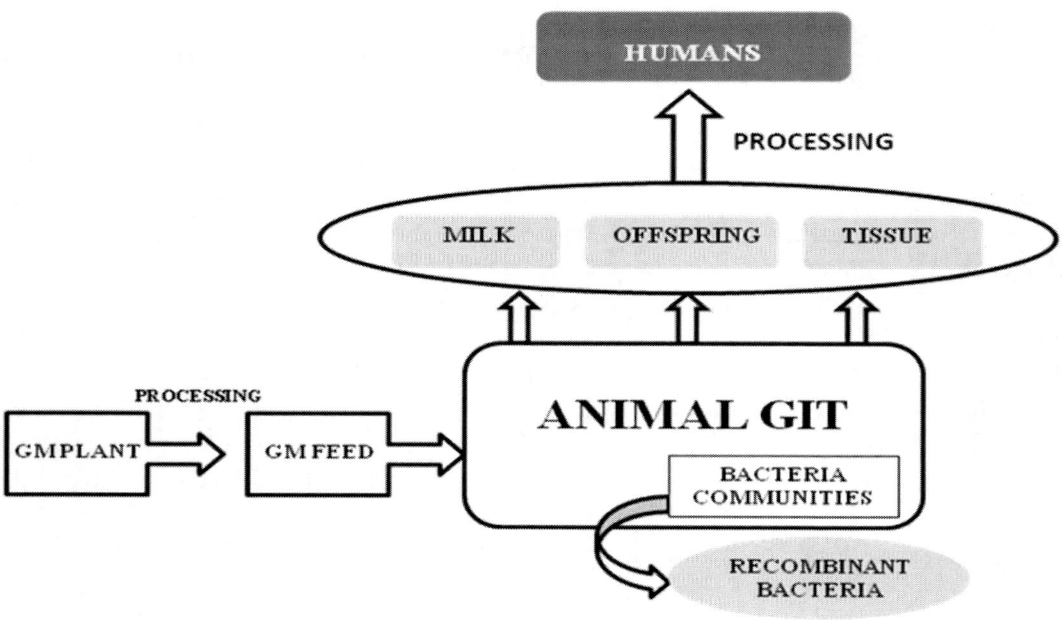

Figure 1. Fate of GM DNA from GM plant to humans through food chain.

To ensure the safety of GM plants as animal feed, regulatory bodies have adopted the policy of substantial equivalence. Though this does not prove nutritional equivalence, to date there have been no adverse effects in animals consuming commercialized GM crops approved on the basis of substantial equivalence. Studies undertaken to address concerns that transgenic protein or DNA may enter the market by means of animal products have shown that recombinant materials in transgenic feed are unlikely to be incorporated into animal products at significant levels. Absorption of plant DNA across the intestinal barrier in food animals does seem to be a normal occurrence when DNA fragments are present at high concentrations. And in general absorption of DNA is a common digestion process and does not appear to have adverse effects (Duggan et al., 2003).

While it might be expected that DNA digestibility would increase with the degree of grinding of the feed, the amount of transgenic DNA ingested will also affect the likelihood of gene persistence within the digestive tract.

The principal conclusion from these findings is that transgenic DNA acts similarly to endogenous DNA, and if present in enough quantity, it may cross the gastrointestinal barrier. The GM DNA can have therefore multiple hypothetical fates, that are subject of this work and summarized in Figure 1.

3. Detection of Transgenic DNA

In genomic studies the transgenic DNA is usually detected by PCR followed by Southern blot and real-time PCR (Matsuoka et al., 2001; Holden et al., 2003; Germini et al., 2004; Hernandez et al., 2004).

PCR is a method to reproduce and amplify a short selected sequence of DNA or RNA. Since it is a very selective and sensitive technique, the original DNA sequence does not need not be pure or abundant, but it can be a minute quantity even in a complex mixture (Holst-Jensen et al 2003). Moreover, PCR is highly efficient so that untold numbers of copies can be made of the DNA in only a few hours. PCR has many uses, e. g. diagnostic of genetic diseases, paternity or biological relationships tests, DNA fingerprinting, DNA forensics, and finding bacteria and viruses. PCR mimics the nature in copying DNA, using the same molecules: two *primers*, short single-stranded DNA sequences that are synthesized to correspond to the beginning and ending of the DNA stretch to be copied; an enzyme called polymerase that moves along the segment of DNA, reading its code and assembling a copy; and a pile of DNA building blocks that the polymerase needs to make that copy. Three major analytical steps are involved in a PCR: denaturation of the double stranded DNA, annealing of the *primers* with the single-stranded sequence of DNA to be copied, and extension of DNA building blocks complementary to the sequence coupled to the primer by polymerase. These three steps are repeated for 30 or 40 cycles. With one cycle, a single segment of double-stranded DNA template is amplified into two separate pieces of double-stranded DNA. These two pieces are then available for amplification in the next cycle. As the cycles are repeated, more and more copies are generated and the number of copies of the template is increased exponentially.

Real-time polymerase chain reaction, also called *quantitative real time polymerase chain reaction* (qPCR), is a development of the PCR technique, which is used to amplify and simultaneously quantify a targeted one or more specific DNA molecule. The procedure follows the general principle of polymerase chain reaction; its key feature is that the amplified DNA is detected as the reaction progresses in *real time*. This is a new approach compared to standard PCR, where the product of the reaction is detected at its end. Two common methods for detection of products in real-time PCR are: (1) non-specific fluorescent dyes that intercalate with any double-stranded DNA, and (2) sequence-specific DNA probes consisting of oligonucleotides that are labeled with a fluorescent reporter which permits detection only after hybridization of the probe with its complementary DNA target. Real-time PCR is often used in food safety, food spoilage and fermentation and for the microbial risk assessment of water quality and in public health protection (Hubner et al., 2001; Peano et al., 2004).

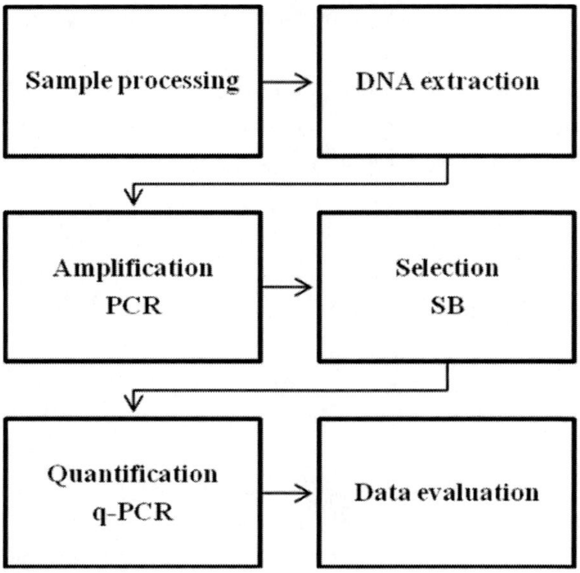

a)

tecnique	PCR	RT-PCR	SB
specificity	+++	++	+
sensitivity	++	+++	+
time	+	++	+
cost	+	++	+++

b)

SB: Southern Blotting; PCR: Polymerase Chain Reaction; RT-PCR: real time PCR.

Figure 2. Methods in GMO detection: (a) analytical process, and (b) features of the techniques.

A Southern blot is a method routinely used in molecular biology for detection of a specific DNA sequence in DNA samples. Southern blotting combines transfer of electrophoresis-separated DNA fragments to a filter membrane and subsequent fragment detection by probe hybridization. Restriction endonucleases are used to cut high-molecular-weight DNA strands into smaller fragments. The DNA fragments are then electrophoresed on an agarose gel to separate them by size. The DNA gel is placed into an alkaline solution (typically containing sodium hydroxide) to denature the double-stranded DNA. A sheet of nitrocellulose membrane is placed and pressured on top of the gel to ensure good and even contact between gel and membrane. Buffer transfer by capillary action from a region of high water potential to a region of low water potential (usually filter paper and paper tissues) is then used to move the DNA from the gel on to the membrane; ion exchange interactions bind the DNA to the membrane due to the negative charge of the DNA and positive charge of the membrane; the membrane is then baked in a vacuum or regular oven to permanently attach the transferred DNA to the membrane; the membrane is then exposed to a hybridization probe, a single DNA fragment with a specific sequence whose presence in the target DNA is to be determined. The probe DNA is labeled so that it can be detected, usually by incorporating a radioactive, fluorescent or chromophoric tag; after hybridization, excess probe is washed from the membrane, and the pattern of hybridization is visualized on X-ray film by autoradiography in the case of a radioactive or fluorescent probe, or by development of color on the membrane if a chromophore has been used. Hybridization of the probe to a specific DNA fragment on the filter membrane indicates that this fragment contains DNA sequence that is complementary to the probe. Southern blots performed with restriction enzyme-digested genomic DNA may be used to determine the number of sequences (e.g., gene copies) in a genome (Van Duijn et al., 2002).

The amount of DNA needed for this technique is dependent on the size and specific activity of the probe. Short probes tend to be more specific.

Under optimal conditions, the LOD is 0.1 pg. Joining these three techniques (PCR, Southern blotting, Real Time PCR) is possible to amplify a DNA fragment, select it between many other sequences and then quantify the specific DNA of interest (see Figure 2.).

Table 1. Occurrence of transgenic DNA (Nd: not detected, X: detected)

Animal	Tissue	Products	Offspring	References
Rodent	x			Grønsberg et al., 2011
Bovine	nd	nd		Einspanier et al., 2001
	nd	x		Nemeth et al., 2004
		x		Agodi et al., 2006
		x		Poms et al., 2003
		x		Castillo et al., 2004
		x		Phipps et al., 2002
	nd			Phipps et al., 2003
Ovine	nd			Tony et al., 2003
	nd			Alexander et al., 2004
	x	x	x	Tudisco et al., 2007
	x			Tudisco et al., 2010
			x	Chowdhury et al., 2004
	x			Sharma et al., 2006
	nd			McAllan & Smith, 1973
Swine	x			Chowdhury et al., 2003a, b
	x			Klotz et al., 2002
	nd			Reuter & Aulrich, 2003
		nd		Weber et al., 2001
	x			Flint & Thomson, 1990
	x			Sharma et al., 2006
	nd			McAllan & Smith, 1973
	x			Mazza et al., 2005
Poultry	x			Chambers et al., 2002
	x	nd		Tony et al., 2003
		x		Einspanier et al., 2001
	x			Deaville and Maddison, 2005
Fish	x			Sanden et al., 2004, 2007
	x	x		Nielsen et al., 2005
	x	x		Chairnak et al., 2008
	nd	nd		Sissener et al., 2010
Humans	x			Martin-Orue et al., 2002
	x			Netherwood et al., 2004

4. Occurrence of Transgenes in Animals.

Identification of DNA fragments in the intestinal wall may constitute a proof of the passage of nucleic acid but it is almost impossible to thoroughly clean the samples from the intestinal content and the risk of false positives due to contamination is too high. Therefore mainly organs were investigated for the presence of diet derived DNA in animal tissues after consumption of GM plants, the results are summarized in Table 1.

4.1. Rodent

Grønsberg et al. detected a 201 bp target sequence in mesenteric lymph nodes, spleen, liver and pancreas from growing rats 2 h after fed on transgenic plant material. After 6 h, target DNA was detectable in the kidneys, and after three days in the liver. Target DNA was not detected in samples from pregnant rats, their foeti or pups. They concluded that low level of feed introduced DNA could be transiently detected in organs of young, growing rats (Grønsberg, 2011). However, indications of increased DNA uptake levels in the GIT of growing rats were not found. Although targeted DNA was more readily detected in the organs of young actively growing rats that the pregnant adults, no indications of a substantially elevated uptake rate in young, growing rats as compared to adults were found.

4.2. Bovine

Because of the highly unstable nature of DNA in the ruminant digestive tract, it is probable that DNA released in the rumen, at least for low copy genes, does not persist to the proximal small intestine. Therefore, passage of plant residues into the ileum prior to their digestion may be necessary for plant genes to have a chance of crossing the intestinal barrier. There is evidence that digestion does occur in the ileum (Alexander et al., 2004) and that plant DNA is released into the aqueous intestinal phase. The first study to probe for plant DNA in ruminant products showed that transgenic DNA was not detectable in the muscle, liver, spleen, kidney, or blood lymphocytes of cattle given ad libitum access to Bt 176 silage (Einspanier et al., 2001). While absorption of plant DNA is possible throughout the ruminant digestive tract, passage of foreign plant low-copy DNA into milk is unlikely. Nemeth searched for DNA fragments of both the high copy number chloroplast-encoded maize rubisco gene (rbcL) and single copy nuclear-encoded transgenic elements (p35S and a MON 810-specific gene fragment) chloroplast DNA fragment in muscle samples from beef steers, broiler chickens, and swine, and in milk samples from dairy cows (Nemeth et al., 2004). Transgenic DNA fragments were not detected in any of the tested animal tissue samples, while fragments of the multicopy chloroplast gene was found in 5, 15, 53% of meat samples respectively, and in 86% of the milk samples. These results suggest that the gene copy number might be a significant factor influencing the detectability of plant DNA sequences in animal tissues. The presence of DNA fragments of high copy number plant sequences in some animal tissues appears to be a normal metabolic occurrence that is variable between different animal species and between tissues types within animal species (Tudisco et al., 2006a).

Einspanier et al. (2004) used real time PCR to quantify transgenic and endogenous genes throughout the digestive tract of cattle fed diets containing Bt 176 maize silage. The amounts of both transgene and endogenous gene increased after passage from the rumen to the abomasum, before decreasing dramatically to unquantifiable levels in the jejunum and colon. DNA is quickly degraded to oligonucleotides, nucleosides and bases throughout the digestive tract: deoxyribonuclease activity has been shown to be present in bovine ruminal fluid and ovine intestinal fluid (Flint and Thomson, 1990; Duggan et al., 2000).

4.2.1 Milk

Agodi and colleagues performed a study on milk, developing a valid protocol by PCR and multicomponent analysis for the detection of specific DNA sequences in milk, focused on GM maize and GM soybean, in order to determine the presence of GM DNA sequences in milk samples collected from the Italian market (Agodi et al., 2006). Moreover, they investigated the stability of transgenic DNA during milk processing, demonstrating that the pasteurization is not able to degrade the DNA sequences in spiked milk samples after pasteurization treatment. Results from the screening of 60 samples of 12 different milk brands demonstrated the presence of GM maize sequences in 25% and of GM soybean sequences in 11.7%. The study is the first to report the presence of GM DNA in milk samples collected from the Italian market. A possible explanation of failing of the other experiments is that their positivity to GM may be fecal or airborne contamination or that the diet derived DNA is degraded in the digestive tract so that only very small fragment sizes were present, and/or to the low sensitivity of the employed extraction methods and PCR assays.

An experimental model (Poms et al., 2003) demonstrated that GM DNA sequences from feeds are detectable in milk after aerosol contamination with feed particles. The probability of feed, and the particle contamination under real conditions is affected by the milking technologies employed. As such, the detection of GM sequences in milk may be interpreted either as a biomarker of transgene feeding via fecal contamination with GM DNA, or, if we exclude it accounts for the ingestion of GM feed, it may indicate the presence of GM feed in the vicinity of milking and milk- storage facilities (Castillo et al., 2004). As such the presence of GM sequences in organic milk may indicate poor-quality assurance by the farmers and/or the producers. One more issue is that one theoretical source of alternative contamination could be the environment: B. thuringiensis, the soil microorganism from which the sequence has its natural origin, could be an environmental contaminant of milk. Likewise, Agrobacterium from which the sequence of GM soybean has been isolated could contaminate the environment around milking and milk-storage facilities. So an alternative source of contamination is possibly recognizable in soil bacteria of the natural environment.

To verify if GM DNA accumulate in milk, meat and eggs derived from livestock fed GM crops an experiment was planned in which herbicide tolerant soybean meal replaced gradually and partially in 8 weeks the conventional rapeseed meal of lactating dairy cows (Phipps et al., 2002). All milk samples weekly collected were negative. The results showed that transgenic DNA could not be detected in milk from cows receiving up to 26% of their diet as herbicide (glyphosate)-tolerant soybean meal. Phipps could detect approximately 0.1% of the input DNA in the milk. It therefore follows that greater than 99.9% of the input DNA was destroyed or degraded to fragments below 180 bp and was undetectable in the milk (Phipps et al., 2003). This extensive degradation of DNA is attributed to the aggressive and extensive digestion processes in the dairy cow (Mc Allan and Smith, 1973).

4.3. Ovine

The feed-ingested recombinant DNA was not detected in visceral tissues (liver, kidney) of lambs or in the spleen from pigs. Of note, however, one liver and one kidney sample from the pigs (different animals) were positive for a 278-bp transgenic fragment.

The study carried by Tony et al. confirms that feed-ingested DNA fragments (endogenous and transgenic) do survive to the terminal GI tract and that uptake into gut epithelial tissues does occur (Tony et al., 2003). A very low frequency of transmittance to visceral tissue was confirmed in pigs (Chowdhury, 2003a,b), but not in sheep (Alexander et al., 2004). It is recognized that the low copy number of transgenes in GM feeds is a challenge to their detection in tissues, but there was no evidence to suggest that recombinant DNA would be processed in the gut in any manner different from endogenous feed-ingested genetic material (Jennings et al., 2003a,b). The potential transfer of plant DNA in blood, spleen, liver, kidney and muscle tissues organs of piglets fed for 35 days with diets containing either GM (MON810) or a conventional maize was investigated (Mazza et al., 2005). Three independent DNA isolations were performed on each organ and tested for the presence of three target gene fragments: Zein, a maize multicopy gene used as positive control to identify maize DNA; Cry1A(b), the synthetic single copy gene inserted into the maize line; Shrunken-2, a maize single copy gene used as control to compare the behaviour of a single copy gene with Cry1A(b). Only a small Cry1A(b) transgene fragment was detected in blood, liver, spleen and kidney of the animals raised with the transgenic feed, while the intact gene or its minimal functional unit were never detected. Statistical analysis of the results showed no difference in recovery of positives for the presence of plant DNA between animals raised with the transgenic feed and animals raised with the conventional feed, indicating that DNA transfer may occur independently from the source and the type of the gene. In this case, blood was the tissue with the highest number of positives, followed by organs rich in blood vessels and involved in filtration, such as liver and kidney. The relatively high recovery of plant DNA in spleen suggests that in young animals the immune and the lymphatic systems may play important roles in nucleotide sequences absorption blood is the main tissue involved in the uptake of short DNA fragments since it collects macromolecules directly absorbed by the intestinal epithelium and the cells of the immune system. These results show that DNA fragments that are progressively degraded or diluted before they reach peripheral positions, therefore DNA transfer to muscle tissues seems to be an unlikely event. In fact, only three cases of foreign DNA presence over 432 tests were detected in muscle.

Moreover in several of these samples, they found fragments of the specific soybean single copy gene (lectin). Transgenic fragments were not found in those samples, which were found positive for chloroplast fragments of control groups of either goats or kids. Transgenic fragments were also found in liver, kidney, heart and muscle (Tudisco et al., 2010).

High copy number chloroplast and low copy soybean lectin genes were found in several samples of goat suckling kids whose mother were fed diet containing conventional (control) or transgenic soybean (treated) (Tudisco et al., 2007). Only in treated group, GM fragments of soybean genes were found in liver, kidney, muscle, spleen, heart, and blood. The presence of feed plant DNA in suckling animals samples could suggest the hypothesis of transfer of gene fragments through the milk or during the gestation (Chowdhury et al., 2004).

It can be assumed that actively growing animals, as well as their of pregnant rats, are more likely to take up DNA from the GIT than mature animals due to the high demand for nutrients for tissue and organ development (Tudisco, 2006b).

4.4. Swine

Unlike ruminant studies, the relative stability of free DNA in pig digestive contents has not been reported.

A 199 bp sequence of plant chloroplast DNA was amplifiable from contents of the stomach, duodenum, jejunum, and ileum for 12 h after feeding, although the intensity of each PCR product eventually diminished over time. In contrast, transgenic sequences of the cry1Ab gene (211 and 251 bp) were not detected from any digestive sample at any time point. This suggests that the number of transgene copies was below the LOD. The actual rate of occurrence of such an event as uptake of transgene is likely so low that an extremely large sample population would be required to demonstrate statistical significance.

Klotz et al. (2002) were also unable to detect transgenic DNA in muscle, liver, spleen, lymph nodes and blood from pigs fed diets containing 200–250 g maize/kg. In addition, and again in contrast to Reuter and Aulrich (2003), chloroplast DNA was also undetectable.

These differences between and within each study likely highlight the significance of the number of genes ingested, or perhaps the sensitivity of methods employed. Another factor that will directly affect gene persistence throughout the digestive tract and therefore indirectly affect the chance of passage across the GIT epithelium is the digestibility of the ingested plant species.

Despite detection of the transgene in pork tissue, there was no difference in the growth performance or meat quality between pigs fed diets containing transgenic or conventional maize (Weber et al., 2001).

Recombinant DNA was detectable in the intestinal contents of pigs up to 48 h after the last feeding of a diet containing the transgenic maize, but not in tissues It is therefore possible that DNA detected near the end of the digestive tract was protected within undigested maize residue that was inaccessible to DNA-degrading enzymes (Reuter and Aulrich, 2003). In contrast, conventional plant DNA fragments were detectable in the investigated pig tissues. That suggest first of all that grinding did not cause any significant disruption of plant DNA. Moreover, even if the gut possesses considerable enzymatic activity to ensure extensive degradation of feed nucleic acids (Flint and Thomson, 1990), the results obtained from the DNA extractions visualized by agarose gel electrophoresis reflect that DNA segments up to 21 kbp are sufficiently stable to survive the enzymatic activities of the gut. The fragments were detected for up to 24 h in the digesta samples, while after 48 h gene fragments were detectable in the contents of the small intestine of one pig and in the content of the rectum of another pig. Any other fragment was not detectable in any content of the GIT from pigs slaughtered 72 h after the last feeding, implying that gene fragments either completely passed the GIT or were finally degraded. In these case study the ovary of pigs contained more plant DNA fragments then the remaining tissues, maybe as a result of intense blood flow during maturation of follicles. Sharma and colleagues detected high copy chloroplast-specific DNA fragment (520bp) in all digesta samples, the majority of intestinal tissues, and at least one of each visceral organ sample from sheep and swine (Sharma et al., 2006). On the other hand each of the five transgene fragments (179-527 bp) surveyed was present in at least 27% of ovine large intestinal content samples and at least 33% of porcine cecal content samples. In sheep, transgene fragments were more common in intestinal digesta than in ruminal or abomasal content (McAllan and Smith, 1973). The feed-ingested recombinant DNA was not detected in visceral tissues (liver, kidney) of lambs or in the spleen from pigs. Of note, however, one liver and one kidney sample from the pigs (different animals) were positive for a 278-bp transgenic fragment. Examination of genomic libraries from these tissues yielded no conclusive information regarding integration of the fragment into porcine DNA (Deaville and Maddison, 2005). Mazza et al. (2005) found transgenes in blood, liver, spleen and kidney of piglets fed with Bt maize.

4.5. Poultry

Chambers et al. (2002) investigated the fate of plant DNA throughout the digestive tract of chickens fed insect-resistant maize event 176. PCR-restriction fragment length polymorphism (RFLP) analysis indicated that the beta lactamase (bla) gene, present as part of the transgenic construct found in event 176 maize, could be detected in the crop of each bird tested and in the stomach of half of them. Results for the transgenic DNA were negative in the small intestine, large intestine, cecum and rectum.

Tony et al.(2003) detected DNA throughout the GIT of poultry fed with Bt176 maize. Using real time PCR, the authors showed that sequences of the maize-specific hmg gene (79 bp) and the transgenic cry1Ab gene (129 bp) were detected in proventriculus, gizzard, duodenum, jejunum, ileum, cecum and rectum; quantitative data were not presented from these studies it appears that short fragments of low copy plant genes have the potential to be absorbed throughout the avian digestive tract and incorporated into body tissues. In contrast, the DNA fragments were not detected in the embryos of eggs. Einspanier et al. (2001), were also unable to detect the chloroplast fragment in the eggs from chickens These studies suggest that plant DNA is less likely to cross into the eggs than to be incorporated into body tissues of chickens. Deaville and Maddison (2005) identified transgenes in the small and large intestine of broilers fed Bt-maize grain and glyphosate tolerant soybean.

4.6. Fish

Nearly one third of the fish consumed by humans is the product of aquaculture. Consequently the needs of commercial feed for intensively cultivating fish are increasing. Most fish prepared feeds use GM soybean meal as good quality plant protein, that have demonstrated the equivalence of nutritional performance (Satoh and Chairnak, 2010). Several studies have investigate the possible transfer of GM DNA from GM soybean feed to fish.

Some small fragments of recombinant DNA have been detected in various gastrointestinal tracts (Sanden et al., 2004; 2007), and in muscle, kidney and blood (Nielsen et al., 2005) of GMO fed Salmon and Rainbow Trout (Chairnak et al., 2008), while Sissener et al. (2010) were not able to find no transgenes in Zebrafish. But after a withdrawal period of five days, where the diet was changed into a non GM feed, no fragments were detected, and no gene expression initiated by the promoter derived from GM soybean was observed in fish cells, suggesting the authors the safety of GM soybean as feed ingredient (Chairnak et al., 2008). Interestingly, often in the studies reported , were detected naturally occurring multicopy plant genes in blood, liver, muscles of various animals. These results suggest that foreign DNA fragments were not completely degraded in the stomach and intestine and might taken up into organs via the GI tracts.

4.7. Humans

Once demonstrating the effective potentiality of foreign DNA transfer from GM feed to livestock and then to food of animal origin, the questions arising regards the passage to human and the effects on animal and human cells. Few studies have been conducted on the fate of foreign DNA ingested by humans. Some studies conducted in vitro indicated that up to 5% of DNA fragments of GM soybean survive a simulated intestinal digestion (Martin-Orue et al., 2002). Studies in vivo revealed the persistence of 4% of GM soya recombinant DNA in ileostomists, while no positive results were obtained in humans with intact GIT, suggesting a significant degradation in the intestine (Netherwood et al., 2004).

Forsman et al. (2003) found fragments of mitochondrial DNA and rabbit transposon RERV-H in the blood of two volunteers who had ingested cooked rabbit meat, in a fraction equivalent of 10^{-8} of the ingested recombinant DNA.

5. R-DNA IN PROCESSED FOOD

5.1. Food Processing Treatments and R-DNA Stability

Many different treatments affect the persistence of DNA in the final food. Heating processes leads to the degradation of intact DNA. However, this does not necessarily render PCR amplification impossible. In an experiment, in which canola meal was pelleted at 100°C, detection of a 1363-bp fragment remained possible (Alexander et al. 2002)

Heating of DNA solutions is the easiest way of examining its effect on DNA fragmentation, by varying processing time and temperature. Hupfer et al. have shown that after 60 min of heat treatment at 95°C the average DNA fragment length in a DNA solution is reduced to less than 600 bp. Similarly, Debode et al. observed that the mean size of the DNA segments is still about 400 bp after the heating of DNA solutions at 99°C for 7 h in a heating block. Nevertheless, even temperatures lower than the cooking temperature may result in degradation of plant DNA. Time of the treatment will also be important, because the longer the heat treatment, the greater the fragmentation of the DNA.

Temperature processing results in the physical degradation of DNA (denaturation). The mechanism of DNA destruction by heat is based on depurination or deamination. Although at temperatures above 100°C significant strand scission and irreversible loss of secondary structure occurs, normal temperatures used for processes such as canning or even autoclaving at 121°C for 15 min do not destroy all the DNA available for PCR (Lindahl, 1993). DNA becomes sheared into smaller fragments and, therefore, PCR sensitivity will be reduced. Even at 70°C DNA is degraded, as shown during the drying of potato sticks for 2 h

In the course of potato chip and flake production, respectively, plant DNA was strongly degraded by heat treatment at 175°C for 3 min and at 150°C for 1 min, and only DNA fragments of 96 bp were amplified from the DNA isolated from the final products (Bauer et al., 2002).

The pH influence on the DNA stability in processed foods has been extensively investigated. Under strongly acidic (pH 3) conditions, depurination of DNA can occur, leading to nicks in the DNA strands and unsuccessful PCR. Moreover, the pH decreases as a result of lactic acid fermentation, which accelerates the degradation of DNA (Fernandes et al., 2013).

In contrast, DNA is relatively stable at alkaline pH. At pH 8.5–9.5, the double strands, of the DNA molecule are separated (Hupefer et al., 1998). Furthermore, in the fermented soy products natto (fermented soybeans) and soy sauce genomic DNA is degraded during the fermentation process (Kakihara et al., 2007, Ogasawara et al., 2003).

Extension of shelf-life and improvement of technological qualities are the objectives of radiation processing of foods. Villavicencio et al. irradiated genetically modified soybeans up to 1000 Gy. DNA damage increased with increasing radiation doses.

A study on the fate of recombinant DNA from genetically modified starter organisms in thermally treated fermented sausages demonstrated how rDNA is protected by the meat matrix against the action of DNase and could be detected during storage of the sausage for 1 9 weeks. The free rDNA represents only a minute fraction of the total rDNA present in the thermally treated food matrix. By far the major part of the rDNA remains entrapped in the killed host cells and is therefore even better protected against degradation. The detectability of rDNA is not only affected by the extent of degradation of the target DNA but also by the choice of primers and the reaction conditions employed in the PCR system (Straub et al., 1999).

5.2. Limitations in R-DNA Detection in Processed Food

Composed products consist of several ingredients with different properties. The nature of each ingredient will determine its extraction efficiency (e.g. oil versus flour) which may alter during processing. As laboratories rarely receive complete compositional data on the products, the choice of a specific extraction technique is usually based on previous experience with similar samples. Furthermore, food products often contain different components derived from the same species, which will be regarded as a single ingredient in the framework of the European 1830/2009/EC GMO labeling rules. With each tissue having a different DNA contribution, and different levels of DNA degradation and/or a depletion mixed in variable unknown ratios, precise quantitative analysis is impossible (Tilley, 2004).

Huge efforts are underway to develop and officially validate tools for GMO analysis. Polymerase chain reaction (PCR) and real-time PCR have become the methods of choice for GMO detection and quantification. It is possible to extract high-quality DNA for GMO analysis from raw materials like seeds or from Certified Reference Materials (CRMs). Unfortunately, many processed food and feed products are often non-optimal sources of DNA: food processing procedures often result in DNA fragments as small as (or even smaller than) ~200–1,000-bp fragments. Moreover, in several refined products, such as glucose syrup, oil or soy lecithin, DNA is depleted to vanishingly small quantities. And, in several complex food products (e.g., sausages or chocolate), DNA may be embedded in substances that inhibit PCR reactions. Thus, overall, the more products are processed and complex, the poorer the quality of the extracted DNA. Several factors influence the applicability and reliability of quantitative PCR analysis on processed samples. First, DNA degradation depletes amplifiable DNA sequences from the analytical sample. This influences the limit of detection (LOD) and the limit of quantification (LOQ) of a method, in that fewer intact target sequences are available as a consequence of DNA fragmentation. This problem can usually be overcome by designing specific primer-probe systems targeting short marker sequences ~60–150 bp in length (Weighardt, 2007; Gryson, 2010)

In order to enhance the sensitivity of real-time PCR and consequently expand the number of analyzable target genes, a preamplification technique has been recently applied to processed foods where DNA can be present in low amounts and/or in degraded forms thereby affecting the reliability of qualitative and quantitative results (Del Gaudio et al., 2010).

6. Expression of Foreign DNA

The main risk for an animal or human being with dietary DNA released during digestion is its integration in the host cells or by the bacteria in the GIT, and subsequent integration in the genome and expression of the foreign genetic information.

Several mechanisms of defence have been developed by higher organisms to avoid penetration of foreign DNA. Among them, cell receptors that can distinguish prokaryotic DNA from the host vertebrate DNA inducing an immunitary response, and nucleases that hydrolyze DNA not associated with DNA –binding proteins, have been described in blood (Zelenay et al., 2003).

Transgene fragment uptake in intestinal epithelial cells has been demonstrated in several studies in many different species in vivo and in vitro, as described in the previous paragraph.

But the hypothetical integration in the host genome and the expression of the foreign DNA is something to be clarified.

Interesting studies have demonstrated the transcriptional activity of the cauliflower mosaic virus 35S CAMV promoter, which is common used in GM plant producing, in human and rodent cells (Myhre et al., 2006).

The risk of integration and expression of foreign DNA appears more concrete in bacteria, for their proper biological characteristics. In fact the acquisition of foreign DNA by horizontal gene transfer (HGT) is recognized as an important way to acquire new genetic evolution and therefore important for bacteria evolution.

The uptake of DNA of variable size in marine bacteria was examined by Jørgensen and Jacobsen (1996). This study reported that uptake of smaller fragments (100 and 250 bp) occurred more rapidly than uptake of larger ones (569 bp). DNA fragments were incorporated into cell constituents as measured by radioactive labeling of the added DNA.

The knowledge of the capacity and mechanism of bacteria transformation in vivo is scarce. There are specific receptor on bacteria membrane for DNA binding, and 20-50 genes are involved in the uptake. It appears as a common trend in many researches the need a similarity between the DNA recombining molecules of 20 bp at least, while only some bacterial species require specific recognition sequence to acquire DNA:(Rizzi et al., 2013). For this reason the possibility of recombination between plant and bacteria seems remote.

DNA fragmentation can lead to a reduced likelihood of uptake in bacteria or to complete genetic inactivation of the DNA. This is interesting in terms of putative spread of GMO-derived DNA, where primarily gene-sized DNA fragments may be relevant. Highly degraded DNA fragments are unlikely to transfer new protein encoding capabilities. However, they may affect allele changes by introducing amino acid substitutions or indels into homologous stretches of a recipient bacterial genome (Nielsen et al., 2007).

Rizzi et al. review about 25 papers dealing with the horizontal transfer of recombinant plant DNA to bacteria, the results were all negative for DNA uptake, except in the presence of homologous regions in both recombinant DNA plant and bacterium (EFSA, 2009, Rizzi et al., 2008). Till now, the homology-based DNA integration research approach has been not explored in deep. The few studied reported on HGT in the lower GIT bacteria, failed to detect any transformation in rats (Wilcks et al., 2004, Kharazmi et al. 2003, Nordgard et al. 2007) and humans (Netherwood et al., 2004) in vivo but gave positive results in rats in vitro (Wilcks et al., 2004).

This results, however, besides being in a small number and not exhaustive, do not treat the problem of the long term effect, in fact recombinant bacteria may take years to multiply to numbers so to be detected by PCR.

Another concern regarding transgenic plant DNA is the possible transfer of antibiotic-resistance markers (ARMs) to bacteria, a dramatic risk highlighted also by health and safety organizations (EFSA, 2009, WHO, 2007). Regarding the use of GM organisms with this modification the cost effectiveness principle is commonly applied: in front of rare bacterial transformation, are privileged the great economical benefits deriving from their use in agriculture. During the approval procedure every new GM organism with this modification must be carefully evaluated case by case, as suggested by the Directive 2001/18/EC (EC, 2001). Furthermore, an antibiotic resistance marker genes are in general in the size range of 700-1000 bp, and therefore difficult to find intact after digestion.

The use of GM plants containing viral inserts (as transgenes or promoters) raises the possibility that these inserts may lead to the creation of new viruses through recombination (HGT). Viruses commonly evolve by recombination and have been shown to recombine with transgenes which are derived from viruses (Greene and Allison 1994, 1996). The creation of new viruses by recombination with a transgene is therefore a plausible risk (De Zoeten 1991; Hull 1994; Gibbs 1994; Allison et al., 1997) that would have consequences on agricultural systems, human health and environmental protection, as many viruses are only controlled by time-consuming precautions and extensive pesticide use (Rybicki and Pietersen 1999; Morales and Anderson 2001). As an example, it has been suggested by Ho that the CaMV promoter, which is a regulatory sequence common to most registered GM plants, could cause cancer through over-expression of oncogenes; the promoter could be integrated into human cells through recombinant events after absorption by GMO food consumption (Ho et al., 1999). On the other hand, it known that CaMV is ubiquitous and its promoter has been detected in food that does not contain transgenic DNA (Wolf et al., 2000).

CONCLUSION

Detection of GMOs in food is an important issue for all the subjects involved in food control and costumer's right. Due to the increasing number of GMOs imported in Europe during the past few years, it has become necessary to screen the products in order to determine the identity of the consumed daily food stuffs.

The problem of the fate of recombinant DNA deriving from GM feed is still an open question. First of all should be reconsidered the GMO labeling system for the food of animal origin to provide the consumer transparent information regarding the use of GM feed, in order to make the citizen able to make a proper choice of what to buy and eat.

This need is more urgent for the lack of conclusive scientific data on the transfer of recombinant DNA from recombinant DNA plants and bacteria along the food chain. Most of the reported studies limitations derive from the severe inadequacy of current analytical methods for the detection of so low frequency events.

This problem is more evident in the worldwide emerging processed food sector. In fact, processed food are often a mixture of various ingredients obtained by a large number of complex processing steps, which contribute to a particular environment where DNA undergoes deterioration and proteins are denatured. The degree of sample processing and the DNA extraction method used may lead to DNA fragment damage, which can be caused by exposure to heat or pressure, enzymatic degradation by nucleases, chemical agents, water activity, shear forces and pH values. All these procedures can affect DNA recovery and the efficiency of PCR current standard protocols and, therefore, the quality of analytical results, leading to under- or overestimation of GMO content.

These evidences open the question of the establishment of standardized international protocols for the quantitative analysis of GMO in foods, validation studies cannot be restricted to the assessment of trueness and precision in (unprocessed) certified reference materials, since such performance parameters may not be unconditionally transferable to results obtained from the analysis of composed and/or processed samples.

REFERENCES

Agodi, A., Barchitta, M., Grillo, A., Sciacca S. (2006). "Detection of genetically modified DNA sequences in milk from the Italian market", *Int. J. Hyg. Environ.-Health* 209:81-88.

Alexander, T.W., Sharma, R., Deng, M.Y., Whetsell, A.J., Jennings, J.C., Wang, Y., Okine, E., Damgaard, D., McAllister T.A. (2004). "Use of quantitative real-time and conventional PCR to assess the stability of the cp4 epsps transgene from Roundup Ready canola in the intestinal, ruminal, and fecal contents of sheep" *J. Biotechnol.* 112:255–266.

Alexander TW, Sharma R, Okine EK, Dixon WT, Forster RJ, Stanford K, McAllister TA (2002) Impact of feed processing and mixed ruminal culture on the fate of recombinant EPSP synthase and endogenous canola plant DNA. *Fems Microbiol Lett* 214:263–269

Allison RF, Schneider WL, Greene AE (1996) Recombination in plants expressing viral transgenes. *Semin Virol* 7: 417-422

Bauer T, Weller P, Hammes WP, Hertel C (2003) The effect of processing parameters on DNA degradation in food. *Eur Food Res Technol* 217(4):338–343

Belloque J., Garcia M.C., Torre M., Marina M.L (2002) Analysis of soybean proteins in meat products: a review. *Critical Reviews in Food Science and Nutrition*, 42: 507–532.

Beever, D. E., Kemp F. (2000). Safety issues associated with the DNA in animal feed derived from genetically modified crops: A review of scientific and regulatory procedures. *Nutr. Abstr. Rev*. 70:197–204.

Beever D. E. and Phipps R. H. (2001). The fate of plant DNA and novel proteins in feeds for farm livestock: A United Kingdom perspective *J. Anim. Sci.* 79: 290–295

Brod F.C.A and Arisi A.C.M (2007), *Recombinant DNA in meat additives: specific detection of Roundup Ready™ soybean by nested PCR Journal of the Science of Food and Agriculture,* 87: 1980–1984.

Castillo, A.R., Gallardo, M.R., Maciel, M., Giordano, J.M., Conti, G.A., Gaggiotti, M.C., Quaino, O., Gianni, C., Hartnell G.F. (2004). Effects of feeding rations with genetically modified whole cottonseed to lactating Holstein cows *J. Dairy Sci*. 87:1778–1785.

Chainark, P., Satoh, S., Hirono, I., Aoki, T., Endo, M. (2008). Availability of genetically modified feed ingredient: Investigations of ingested foreign DNA in rainbow trout Oncorhynchus mykiss. *Fish. Sci*. 74: 380–390.

Chowdhury, E.H., Kuribara, G., Hino, A., Sultana, P., Mikami, O., Shimada, N., Guruge, K.S., Saito, M., and Y. Nakajima. (2003a). Detection of corn intrinsic and recombinant DNA fragments and Cry1Ab protein in the gastrointestinal contents of pigs fed genetically modified corn Bt11. *J. Anim. Sci.* 81: 2546–2551.

Chowdhury, E.H., Mikami, O., Nakajima, Y., Kuribara, M., Suga, K., Hanazumi, M., Yomemochi C.. (2003b). Detection of genetically modified maizeDNAfragments in the intestinal contents of pigs fed StarLink CBH351. *Vet. Hum. Toxicol*. 45:95–96.

Chowdhury, E.H., Mikami, O., Murata, H., Sultana, P., Shimada, N., Yoshioka, M., Guruge, K.S., Yamamoto, S., Miyazaki, S., Yamanaka, N., Nakajima Y. (2004). Fate of maize intrinsic and recombinant genes in calves fed genetically modified maize Bt11. *J. Food Prot*. 67:365–370.

Deaville, E.R., and. Maddison C.B. (2005). Detection of Transgenic and Endogenous Plant DNA Fragments in the Blood, Tissues, and Digesta of Broilers *J. Agric. Food Chem.* 53:10268-10275.

Del Gaudio S., Cirillo A., Di Bernardo G., Galderisi U. , Cipollaro M. (2010). A preamplification approach to GMO detection in processed foods *Anal Bioanal Chem* 396:2135–2142.

Debode F, Janssen E, Berben G (2007) Physical degradation of genomic DNA of soybean flours does not impair relative quantification of its transgenic content. *Eur Food Res Technol* 226:273–280.

de Zoeten, G.A. (1991). Risk assessment: Do we let history repeat itself? *Phytopathology* 21, 585-586.

Duggan P S, Chambers P A., Heritage J, Forbes J. M. (2003). Fate of genetically modified maize DNA in the oral cavity and rumen of sheep *British Journal of Nutrition*, 89, 159–166.

Einspanier, R., Klotz, A., Kraft, J., Aulrich, K., Poser, R., Schw¨agele, F., Jahreis, G., and Flachowsky, G. (2001). The fate of forage plant DNA in farm animals: A collaborative case-study investigating cattle and chicken fed recombinant plant material. *Eur. Food Res. Technol.* 212: 129– 134.

Einspanier R, Lutz B, Rief S, Berezina O, Zverlov V, Schwarz W, Mayer J (2004) Tracing residual recombinant feed molecules during digestion and rumen bacterial diversity in cattle fed transgene maize. *Eur Food Res Technol* 218:269–273.

EFSA. (2007). "Statement on the fate of recombinant DNA or proteins in meat, milk and eggs from animals fed with GM feed" Retrieved February 20, 2012, from *http://efsa.europa.eu/EFSA/efsa_locale.*

EFSA. (2008). GMO Panel Working Group on Animal Feeding Trials. "Safety and nutritional assessment of GM plants and derived food and feed: the role of animal feeding trials" *Food Chem Toxicol* 46(1):S2–S70.

EFSA (2009). Consolidated presentation of the joint Scientific Opinion of the GMO and BIOHAZ Panels on the "Use of antibiotic resistance genes as marker genes in genetically modified plants" and the Scientific Opinion of the GMO Panel on "Consequences of the opinion on the use of antibiotic resistance genes as marker genes in genetically modified plants on previous EFSA assessments of individual GM Plants."*http://www.efsa.europa.eu/en/scdocs/doc/gmo biohaz st ej1108 ConsolidatedARG summary en,0.pdf*

EC Directive 2001/18/EC of the European Parliament and of the Council of 12 March 2001 on the deliberate release into the environment of genetically modified organisms and repealing Council Directive 90/220/EEC. *Official J. Eur. Commun.* L 106: 1–39. *http://europa.eu.int/eurlex/en/lif/dat/2001/en 301L0018.html*

EU Commission (2003) REGULATION (EC) No 1830/2003 OF THE EUROPEAN PARLIAMENT AND OF THE COUNCIL of 22 September 2003 concerning the traceability and labelling of genetically modified organisms and the traceability of food and feed products produced from genetically modified organisms and amending Directive 2001/18/EC *Official Journal of the European Union* 18.10.2003

Eurobarometer. (2010). *Risk Perception Report, Brussels.*

Eurostat. (2011). "The food chain for animal and vegetable fats and oils. A statistical portrait based on 2005-2010 statistics" *Brussels: Statistic in Focus* 62/2011.

Flint, H.J., and Thomson A.M.(1990). Deoxyribonuclease activity in rumen bacteria. *Lett. Appl. Microbiol.* 11:18–21.

FAO/WHO (2000). Safety aspects of genetically modified foods of plant origin. Report of a Joint FAO/WHO Expert Consultation on Foods Derived from Biotechnology, World Health Organization (WHO), Geneva, Switzerland, p. 35. *http://www.who.int/foodsafety/publications/* biotech/en/ec june2000 en.pdf.

Fernanades TJR, Oliveira MBPP, Mafra I (2013). Tarcing transgenic maize as affected by breadmaking process and raw material for the production of a traditional maize bread, broa. *Food Chemistry*, 138: 687-692.

Forsman, A., Ushameckis, D., Bindra, A., Yun Z., Blomberg, J. (2003).Uptake of amplifiable fragments of retrotransposon DNA from the human alimentary tract. *Mol. Genet. Genomics* 270: 362–368.

Germini, A., Zanetti, A., Salati, C., Rossi, S., Forre, C., Schmid, S.,. Marchelli R. (2004). Development of a seven target multiplex PCR for the simultaneous detection of transgenic soybean and maize in feeds and foods. *J.Agric Food Chem* 52:3275–3280.

Gibbs M (1994) Risks in using transgenic plants? *Science* 264: 1650-1651.

Greene AE, Allison RF. (1994). Recombination between viral RNA and transgenic plant transcripts. *Science* 263: 1423-1425.

Greene AE, Allison RF. (1996). Deletions in the 3' untranslated region of cowpea chlorotic mottle virus transgene reduce recovery of recombinant viruses in transgenic plants. *Virology* 225: 231-234.

Greiner, R. and Konietzny U., (2008). Presence of genetically modified maize and soy in food products sold commercially in Brazil from 2000 to 2005. *Food Control*, 19: 499-505.

Gryson N. (2010). Effect of food processing on plant DNA degradation and PCR-based GMO analysis: a review *Anal Bioanal Chem* 396:2003–2022.

Grønsberg L., Idun M., Nordgard, L., Fenton, K., Hegge, B., Nielsen, Kaare M., Bardocz, S., Pusztai, A., Traavik. T. (2011). Uptake and organ distribution of feed introduced plasmid DNA in growing or pregnant rats. *Food Nutr Sci* 2: 377-386.

Hernandez, M., Duplan, M.-N., Berthier, G., Vaitilingom, M., Hauser,W., Freyer, R., Pla, M., and Y. Bertheau. (2004). Development and comparison of four real-time polymerase chain reaction systems for specific detection and quantification of Zea mays. *J Agric Food Chem* 52:4632–4637.

Ho M-W et al. (1999). Cauliflower mosaic viral promotor-a recipe for disaster? *Microb. Ecol. Health Dis.* 11:194-97

Holden, M.J., Blasic, J.R., Bussjaeger, L., Kao, C., Shokere, L.A., Kendall, D.C., Freese, L., Jenkins G.R.. (2003). Evaluation of extraction methodologies for corn kernel (Zea mays) DNA for the detection of trace amounts of biotechnology-derived DNA. *J Agric Food Chem* 51:2468–2474.

Holst-Jensen, A., Ronning, S.B., Lovseth, A., Berdal. K.G. (2003). PCR technology for screening and quantification of genetically modified organisms (GMOs). *Anal Bioanal Chem* 375:985–993.

Hubner, P., Waiblinger, H.-U., Pietsch, K., Brodmann P.(2001). Validation of PCR methods for quantitation of genetically modified plants in food *J AOAC Int* 84:1855–1864.

Hupfer C, Hotzel H, Sachse K, Engel KH (1998) *Detection of the genetic modification in heat-treated products of Bt maize by polymerase chain reaction.* Z Lebensm Unters Forsch 206 (3):203–207

Hull, R.(1994) Resistance to plant viruses: obtaining genes by non conventional approaches. *Euphytica* 75 195-205.

Hupfer C, Mayer J, Hotzel H, Sachse K & Engel KH (1999) The effect of ensiling on PCR-based detection of genetically modified Bt maize. *European Food Research and Technology* 209, 301–304.

Hupfer C, Hotzel H, Sachse K, Moreano F, Engel KH (2000) PCR-based quantification of genetically modified Bt maize: single-competitive versus dual-competitive approach. *Eur Food Res Technol* 212(1):95–99

James C., (2012). Global Status of Commercialized Biotech/GM Crops: 2011. ISAAA Brief No. 44. ISAAA: Ithaca, NY.

Jennings, J.C., Kolwyck, D.C., Kays, S.B., Whetsell, A.J., Surber, J.B., Cromwell, G.L., Lirette, R.P., and. Glenn K.C (2003a). Determining whether transgenic and endogenous plant DNA and transgenic protein are detectable in muscle from swine fed Roundup Ready soybean meal *J Anim Sci* 81:1447–1455.

Jennings, J.C., Whetsell, A.J., Nicholas, N.R., Sweeney, B.M., Klaften, M.B., Kays, S.B., Hartnell, G.F., Lirette, R.P., Glenn K.C.(2003b). Determining whether transgenic or endogenous plant DNA is detectable in dairy milk or beef organs. *Bull. Int. Dairy Fed.* 383, 41–46.

Jørgensen NOG, Jacobsen CS (1996) Bacterial uptake and utilization of dissolved DNA. *Aquat. Microb. Ecol.* 11: 263–270.

Kakihara Y, Matsufuji H, Chino M, Yamagata K (2007). Detection of recombinant DNA of genetically modified (GM) soybeans in heat-treated GM soybeans and commercial natto. *Food Control* 18(10):1289–1294.

Kharazmi, M., Sczesny, S., Blaut, M., Hammes, W. P., Hertel, C. (2003). Marker rescue studies of the transfer of recombinant DNA to Streptococcus gordonii in vitro, in foods, and gnotobiotic rats. *Appl. Environ. Microbiol.* 69: 6121–6127.

Klotz, A., J. Meyer, Einspanier R.. (2002). Degradation and possible carry over effects of feed DNA monitored in pigs and poultry. Eur. Food Res. Technol. 214:271–275.

Lindahl T (1993) Instability and decay of the primary structure of DNA. *Nature* 362(6422):709–715

Lin, H. Y., Chiueh, L. C., Shih, D. Y. C. (2000). Detection of genetically modified soybeans and maize by the polymerase chain reaction method. *J. Food Drug Anal.* 8:200-207.

Martin-Orue, S. M., O'Donnell, A. G., Arino, J., Netherwood, T., Gilbert, H. J., and Mathers, J. (2002). Degradation of transgenic DNA from genetically modified soya and maize in human intestinal simulations. *Br. J. Nutr.* 87: 533–542.

Matsuoka, T., Kuribara, H., Akiyama, H., Miura, H., Goda, Y., Kusakabe, Y., Isshiki, K., Toyoda, M., Hino A.. (2001). A multiplex PCR method of detecting recombinant DNAs from five lines of genetically modified maize. *J Food Hyg Soc Jpn* 42:24–32.

Mazza, R., Soave, M., Morlacchini, M., Piva, G., Marocco A.(2005). Assessing the transfer of genetically modified DNA from feed to animal tissues. *Transgenic Res* 14:775–784.

Morales FJ and Anderson P (2001) The emergence and dissemination of whitefly-transmitted geminiviruses in Latin America. *Arch. Virol.* 146: 415-441.

McAllan, A.B., and Smith R.H. (1973). Degradation of nucleic acids in the rumen. *Br J Nutr* 29:331–345.

McAllan, A. B. (1982). The fate of nucleic acids in ruminants. *Proc. Nutr. Soc.* 41:309–317.

Nemeth, A., Wurz, A., Artim, L., Charlton, S., Dana, G., Glenn, K., Hunst, P., Jennings, J., Shilito, R., Song P. (2004). Sensitive PCR analysis of animal tissue samples for fragments of endogenous and transgenic plant DNA. *J Agr Food Chem* 52:6129–6135.

Netherwood, T., Martin-Orue, S., O'Donnell, A. G., Gockling, S., Graham, J., Mathers, J. C., Gilbert, H. J. (2004). Assessing the survival of transgenic plant DNA in the human gastrointestinal tract. *Nature Biotechnol.* 22:204–209.

Myhre, R. M., Fenton, K. A., Eggert, J., Nielsen, K. M., and Traavik, T. (2006).The 35S CaMV plant virus promoter is active in human enterocyte-like cells.Eur. *Food Res. Technol.* 222: 185–193.

Netherwood, T., Martin-Orue, S., O'Donnell, A. G., Gockling, S., Graham, J., Mathers, J. C., Gilbert, H. J. (2004). Assessing the survival of transgenic plant DNA in the human gastrointestinal tract. *Nature Biotechnol.* 22: 204–209.

Nielsen, C. R., Berdal, K. G., Bakke-McKellup, A. M., Holst-Jensen, A. (2005). Dietary DNA in blood and organs of Atlantic salmon (Salmo salar L.). *Eur. Food Res. Technol.* 221: 1–8.

Nielsen KM, Johnsen P, van Elsas JD (2007) Gene transfer and micro-evolution in soil. In van Elsas JD, Janssen JK, Trevors J, eds, *Modern Soil Microbiology*, 2nd edn, CRC Press, pp 55–81.

Nordgard, L., Nguyen, T., Midtvedt, T., Benno, Y., Traavik, T., Nielsen, K. M. (2007). Lack of detectable DNA uptake by bacterial gut isolates grown in vitro and by Acinetobacter baylyi colonizing rodents in vivo. *Environ. Biosafety Res.* 6: 149–160.

Ogasawara T, Arakawa F, Akiyama H, Goda Y, Ozeki Y (2003) Fragmentation of DNAs of processed foods made from genetically modified soybeans. *Jpn J Food Chem* 10(3):155–160.

Peano, C., Samson M.C., Palmieri, L., Gulli, M., Nelson M. (2004). Qualitative and quantitative evaluation of the genomic DNA extracted from GMO and non-GMO foodstuffs with four different extraction methods. *J Agric Food Chem* 52:6962–6968.

Phipps R.H., Beever D.E., Humphries D.J.. (2002).Detection of transgenic DNA in milk from cows receiving herbicide tolerant (CP4EPSPS) soyabean meal. *Livest Prod Sci* 74:269–273.

Phipps R.H., Deaville E.R., Maddison B.C. (2003). Detection of transgenic and endogenous plant DNA in rumen fluid, duodenal digesta, milk, blood, and feces of lactating dairy cows. *J Dairy Sci* 86: 4070–4078.

Poms R.E, Hochsteiner W., Luger K., Glossl J., Foissy H. (2003). Model studies on the detectability of genetically modified feeds in milk. *J. Food Prot.* 66. 304–310.

Reuter, T. and Aulrich K. (2003). Investigations on genetically modified maize (Bt-maize) in pig nutrition: fate of feed-ingested foreign DNA in pig bodies *Eur Food Res and Technol* 216:185–192.

Rybicki E and Pietersen G (1999). Plant virus disease problems in the developing world. *Adv. Virus. Res.* 53:127-164.

Rizzi, A., Brusetti, L., Arioli, S., Nielsen, K. M., Tamagnini, I., Tamburini, A., Sorlini, C., and Daffonchio, D. (2008). Detection of feed-derived maize DNA in goat milk and evaluation of the potential of horizontal transfer to bacteria. *Eur. Food Res. Technol.* 227: 1699–1709.

Rizzi A , Raddadi N , Sorlini C , Nordgrd L, Nielsen KM, Daffonchio D. (2012). The Stability and Degradation of Dietary DNA in the Gastrointestinal Tract of Mammals: Implications for Horizontal Gene Transfer and the Biosafety of GMOs. *Critical Reviews in Food Science and Nutrition,* 52:142–161.

Sanden, M., Ian, J.B., Rahman, M A., Hemre, GI. (2004). The fate of transgenic sequences present in genetically modified plant products in fish feed, investigating the survival ofGM soybean DNA fragments during feeding trials in Atlantic salmon, Salmo salar L. *Aquaculture*. 237: 391–405.

Sanden, M., Berntssen, M.H.G., Hemre, G.I. (2007). Intracellular localization of dietary and naked DNA in intestinal tissue of Atlantic salmon, Salmo salar L. using in situ hybridization. *Eur. Food Res. Technol.* 225: 533–543.

Satoh S. and Chairnak P (2010). Availability of genetically modified ingredients for rainbow trout. *Bulletin Fishing Research Agency* 31: 63-69.

Sharma, R., Damgaard, D., Alexander, T.W., Dugan, M.E.R., Aalhus, J.L., Stanford, K., McAllister T.A.. (2006). Detection of transgenic and endogenous plant DNA in digesta and tissues of sheep and pigs fed Roundup Ready canola meal. *J Agr Food Chem* 54:1699–1709.

Sissener, N. H., Johannessen, L. E., Hevrøy, E. M.,Wiik-Nielsen, C. R., Berdal, K. G., Nordgreen, A., and Hemre, G.-I. (2010). Zebrafish (Danio rerio) as a model for investigating the safety of GM feed ingredients (soya and maize); performance, stress response and uptake of dietary DNA sequences. *British J. Nutr.* 103: 3–15.

Straub C., Joachim A, Hertel W. P. H (1999). The fate of recombinant DNA in thermally treated fermented sausages *Eur Food Res Technol* 210 :62–67.

Tony, M.A., Butschke, A., Broll, H., Grohmann, L., Zagon, J., Halle, I., Danicke, S., Schauzu, M., Hafez, H.M., Flachowsky G. (2003). Safety assessment of Bt 176 maize in broiler nutrition: degradation of maize-DNA and its metabolic fate. *Arch Anim Nutr* 57:235–252.

Tilley M (2004). PCR amplification of wheat sequences from DNA extracted during milling and baking. *Cereal Chem* 81 (1):44–47.

Tudisco, R., Infascelli, F., Cutrignelli, M.I., Bovera, F., Morcia, C., Faccioli, P., Terzi V. (2006a). Fate of feed plant DNA monitored in water buffalo (Bubalus bubalis) and rabbit (Oryctolagus cuniculus). *Livest Sci* 105:12–18.

Tudisco, R., Lombardi, P., Bovera, F., D'Angelo, D., Cutrignelli, M.I., Mastellone, V., Terzi, V., Avallone, L., Infascelli F. (2006 b). Genetically modified soybean in rabbit feeding: detection of DNA fragments and evaluation of metabolic effects by enzymatic analysis. *Anim Sci* 82:193–199.

Tudisco, R., Cutrignelli, M.I., Calabrò, S., Guglielmelli, A., Infascelli F. (2007). Investigation on genetically modified soybean (RoundUp Ready) in goat nutrition: DNA detection in suckling kids. *Ital J Anim Sci* 6(1):380-382.

Tudisco, R. et al. (2010). Fate of transgenic DNA and evaluation of metabolic effects in goats fed genetically modified soybean and in their offsprings. *Animal*, 1-10.

Van Duijn, G., Van Biert, R., Bleeker-Marcelis, H., Van Boeijen, I., Jama Adan, A., Jhakrie, S., Hessing M. (2002). Detection of genetically modified organisms in foods by protein- and DNA-based techniques: bridging the methods. *J AOAC Int* 85:787–791.

Villavicencio A, Araujo MM, Baldasso JG, Aquino S, Konietzny U, Greiner R (2004) Irradiation influence on the detection of genetic-modified soybeans. *Radiat Phys Chem* 71(1–2):491–494.

Weber, T. E.,. Richert B. T. (2001). Grower-finisher growth performance and carcass characteristics including attempts to detect transgenic plant DNA and protein in muscle from pigs fed genetically modified "Bt" corn. *J. Anim. Sci.* 79 (Suppl. 2):67-73.

Weighardt F. (2007). GMO quantification in processed food and feed *Nature Biotechnology* 25, 1213- 1214.

WHO (World Health Organization) (2002), Food Safety Programme, *20 Questions on genetically modified (GM) foods.*

WHO. (World Health Organization) (2007). Critically important antimicrobials for human medicine: Categorization for the development of risk management strategies to contain antimicrobial resistance due to nonhuman antimicrobial use. *Report of the second WHO expert meeting,* Copenhagen, 29–31 May 2007.

Wilcks, A., van Hoek, A. H., Joosten, R. G., Jacobsen, B. B., Aarts, H. J. (2004). Persistence of DNA studied in different ex vivo and in vivo rat models simulating the human gut situation. *Food Chem. Toxicol.* 42: 493–502.

Wolf, C.; Scherzinger, M.; Wurz, A.; Pauli, U.; Hübner, P., Lüthy, J. (2000). Detection of cauliflower mosaic virus by the polymerase chain reaction: testing of food components for false-positive 35S-promoter screening results. *Eur Food Res and Tech.* 210, (5): 367-372.

Zelenay, S., Elıas, F., Flo, J. (2003). Immunostimulatory effects of plasmid DNA and synthetic oligodeoxynucleotides. *Eur J of Imm* 33 (5): 382–1392.

In: Processed Foods
Editors: Chloe M. Gagne and Daniel B. Jones
ISBN: 978-1-62808-688-1

Chapter 3

CASSAVA (*MANIHOT ESCULENTA* CRANTZ) AS A SOURCE OF CHEMICALLY SAFE FOOD: A CRITICAL REVIEW

G. J. B. Gnonlonfin[1,*] and L. Brimer[2]

[1]Biosciences Eastern and Central Africa, International Livestock Research Institute, Old Naivasha Road, Nairobi, Kenya

[2]Department of Veterinary Disease Biology, Faculty of Health and Medical Sciences, University of Copenhagen, Frederiksberg, Denmark

ABSTRACT

The issue of food contaminants is broad ranging from undesirable natural constituents or unauthorized additives over environmental pollutants and to contaminants related to specific steps of food production, such as fermentation, heat treatments or to cross-contamination. In some cases, mainly mycotoxins and toxic natural plant constituents and their derivatives pose a risk for animal and human welfare. Risk assessment issues include the identification and characterization of the toxicological hazards, identification of the possible routes of the contaminants to plant based food and the carry-over of the parent compounds or their metabolites to animal derived foods; all in order to define exposure scenario(s). Based on this situations that may require risk management measures can be pinpointed. Some scientists, regulators, non-governmental organizations and consumers are raising concerns about potential risks related to the increasing use of cassava and its products.

In this paper, we broadly review risks due to chemicals that may be present in cassava (*Manihot esulenta* Crantz) products and provide a brief overview of some of the acceptable limits comparing levels in cassava and other crops. In conclusion, we point out the safety of most cassava products.

Keywords: Cassava, cyanogens, mycotoxins, food safety

---

* Corresponding authors: e-mail address: b.gnonlonfin@cgiar.org or bgnonlonfin@yahoo.fr (G.J.B. Gnonlonfin).

Introduction

Cassava (*Manihot esculenta* Crantz, Euphorbiaceae) is widely grown as a staple food and animal feed in countries of tropical and sub-tropical Latin America from where it originates, Africa and Asia between 30° N and 30° S with a total cultivated area over 13 million hectares, more than 70% of it being in Africa and Asia (FAO, 2000; Westby, 2002). In comparison to others crops cassava counts as one of the most important sources of food. Hence, concerning carbohydrate this commodity, after rice, sugarcane and maize, is of major importance nourishing over 500 million people mainly in the developing countries of the tropics and sub-tropics. Its main value is in its storage roots with a dry matter containing more than 80% of starch (El-Sharkawy, 2003). However, despite its economic importance, this crop suffers from two major disadvantages.

Firstly, all parts of the plant contain cyanogenic glycosides that can release toxic cyanide. A Codex Alimentarius Commission (CAC) standard exists for so-called sweet cassava roots for trade as a commodity allowing unprocessed consumption (snacking). The maximum level of total cyanide (HCN/CN^-) to be potentially released for such a root is 50 mg kg^{-1} fresh weight (CAC, 2003/2005). However, imported roots for sale in retail shops as sweet cassava in Europe have been shown to often have a higher content (Kolind-Hansen and Brimer, 2010), as has frozen peeled large pieces imported to Australia (Burns et al., 2012). Thorough processing of the roots significantly reduces or eliminates this toxicity prior to consumption (El-Tinay et al. 1984; Brimer, 2001; Nyirenda et al. 2011). Such processing is needed for so-called bitter cassava roots, which by a later published CAC standards were directly defined as those releasing more than 50 mg HCN kg-1 fresh weight (CAC, 2010).

Secondly, a major constraint for the use of cassava is that farmers, processors and consumers may suffer substantial losses during storage of the fresh root tubers. This is due to rapid physiological deterioration processes that initiate within 24-48 hours after harvest followed by a second stage (after 5 to 7 days) involving microbial decay (Beeching et al. 1998). During the deterioration the content of hydroxycoumarins such as scopoletin increases (Buschmann et al. 2000; Gnonlonfin et al. 2011).

Nutrients in Cassava Tissues and Organs

The protein content in the roots is low, i.e. in the interval 1.5 to 4.7 % (d.w.) for the mean values of crude protein found in six investigations (OECD, 2009), and may only be slightly enhanced by fermentation (Tivana et al. 2007). The root protein has less than the recommended minimum of all essential amino acids except for tryptophan. The nitrogen-to protein conversion factor based on nitrogen recovered from total amino acid analysis including ammonia ranges from 4.75 to 5.87; the traditional factor of 6.25 being non-valid for cassava root (Yeoh and Truong, 1996).

The leaves are used in salads and in the preparation of relishes. In Congo and Central Africa leaves are fermented to the food product Ntoba mbodi (Mokemiabeka et. al. 2011). The range of mean values from five investigations of the crude protein content is 14.7-36.4 % (d.w.) (OECD, 2009).

In contrast to what is the case for the roots, the leaf levels of all of the essential amino acids except methionine exceed the FAO's recommendations. In the leaves the content of the amino acids glutamic acid and leucine is the highest, with values above 4 g/100 g true protein, followed by aspartic acid, arginine, and alanine (OECD, 2009).

Cassava roots with yellow parenchyma have elevated contents of carotenoids; some with vitamin A effect. Total carotene may vary between 1 and 18 μg/g (f.w.) (Carvalho et al., 2012). Vitamin levels in mature cassava roots and leaves are low, young leaves normally eaten by humans being richer in ascorbic acid, thiamine and β-carotene (OECD, 2009).

FUNGAL CONTAMINATION OF POST-HARVEST ROOTS AND ROOT PRODUCTS

To overcome the post-harvest losses, cassava is usually processed soon after harvesting. One method of processing is to cut and dry roots to make cassava chips especially during the dry season which in West Africa is from November to March. However there is a risk of fungal contamination during the drying period. Infection of cassava chips with fungi, especially species of *Aspergillus*, *Fusarium* and *Penicillum* has been previously reported (Muzanila et al. 2000; Gnonlonfin et al. 2008, 2012b). Fungal contamination can lead to discoloration, mouldy taste, production of off-odours and possibly the production of mycotoxin that are harmful to animals and humans (Shephard, 2008). Of particular concern would be the potential for contamination with aflatoxins, since different cassava products have been shown to be infected by among others *Aspergillus flavus*, a potential producer of these toxins. In general aflatoxins are a commonly occurring group of mycotoxin formed by *Aspergillus* spp in tropical and sub-tropical regions and which has been previously found among others in maize and maize-based products (Fandohan et al. 2005; Matumba et al. 2009).

Aflatoxins occur worldwide in many food commodities and derived food products. Thus, recent investigations in South Korea showed that maize, soybean, nuts and nut products as well as spices were contaminated with aflatoxin B_1 (Chun et al. 2006). As another example it can be mentioned that samples of dry roasted groundnuts purchased from street hawkers, markets and retail shops in southwestern Nigeria were analyzed and found to contain aflatoxin B_1 residues with a mean level of 25.5 μg kg^{-1} (Bankole et al. 2005), while Ediage et al. (2011) recently among 15 samples of peanut cake from Benin analysed found one to be contaminated at a level of 282 μg kg^{-1} aflatoxin B_1. Also, the presence of rather low concentrations of aflatoxins in cassava chips from Congo Brazzaville andTanzania (Manjula et al. 2009) as well as flour from chips in Benin (Ediage et al. 2011) has been reported. However, other studies found no aflatoxin in cassava samples collected from Tanzania (Muzanila et al. 2000), Ghana (Wareing et al. 2001), Nigeria (Jimoh and Kolapo, 2008) and Benin (Gnonlonfin et al. 2008; Gnonlonfin et al. 2012b). In contrast, a few other less toxic mycotoxins such as sterigmatocystin, patulin, cyclopiazonic acid, penicillic acid and tenuazonic acid were reported to be present in cassava product (Wareing et al. 2001). Because of the important role that cassava plays in guaranteeing food security in among others sub-Saharan Africa there is a need to evaluate the overall chemical food safety of the fresh cassava and its products. Thus, the present literature study was conducted to obtain an overview of the prevailing levels of cyanogenic toxins in different products of cassava root tubers, the consequences if any of the occurrences of high concentrations of hydroxycoumarins (scopoletin etc.) and the mycotoxin status with special emphasis on aflatoxins in chips and other products made from different grown cassava cultivars.

CASSAVA AND CYANOGENS

Cyanogenic glycosides belong to the class of secondary plant metabolites that are bio-activated by the action of β-glucosidases upon disruption of plant tissues; a class of metabolites called phytoanticipins. To this group belong also glucosinolates, avenacosides and benzoxazinoid glucosides (Jørgensen et al. 2005; Morant et al. 2008). In cassava we find D-glucose joined by a β-linkage to acetone cyanohydrin i.e. linamarin (Dunstan et al. 1906) which constitutes the major part of the cyanogens naturally present together with lotaustralin (Nartey, 1968); in cassava roots as well as leaves. Additionally, 2-(6-o-(β-D-apiofuranosyl)-β-D-glucopyranoxyl)oxy)-2-methylbutanenitrile have been found in the cassava root cortex (Prawat et al. 1995).

The amounts of glycosides in cassava roots varies from harmless to lethal depending on the variety, soil fertility status or soil moisture condition. Depending on variety the content of cyanogenic glucosides in the edible root parenchyma thus may vary from about 1 and to approximately 6 - 700 mg kg^{-1} (fresh weight basis) when calculated as the total cyanogenic potential; i.e as mg HCN eq kg^{-1} (Bokanga, 1994a; Chiwona-Karltun et al. 2004). Based on the cyanogenic glycosides concentration, cassava plants as already mentioned often are distinguished as the bitter varieties (high cyanogenic glycosides concentration) and sweet (cool) varieties (low in cyanogens) (Dofour, 1993; Chiwona-Karltun et al. 1998; El-Sharkawy, 2003; Mkumbira et al. 2003).

Not only the roots but also the leaves are used as food and feed (Ravindra, 1993; Ndugi et al., 2003). The cyanogenic potential of leaves of the manihot plant is much higher (5-20 times than that of the roots) but is in general less varying (Bokanga, 1994a,b; Ngudi et al. 2003; Kolind-Hansen and Brimer, 2010). The high concentrations of cyanogenic glycosides in the leaves do not normally present a problem for their use in food, however, taking the general methods of preparation into consideration (Bokanga, 1994b).

The actual lethal toxic agent is hydrocyanic acid (HCN) readily released by β-glycosidase catalysed enzymatic hydrolysis of the cyanogenic glycoside during processing and digestion (Herbert, 1988; Brimer, 2001). The view that the acute toxicity in fact may not be entirely due to the cyanogens, has been expressed by workers who based their arguments on the fact that a greater proportion of the HCN content in cassava is lost during processing (Obidoa and Obasi, 1991). This issue will be addressed later.

On several occasions it was reported that a stable diet of cassava in addition to acute poisoning with acutely vomiting, dizziness, constriction of the throat, headaches, nausea and even death in severe cases (Westby, 2002) also may produce human chronic intoxication. The consequences supposedly are among others endemic goiter, cretinism and mental retardation as early summarized by Dorozynsky (1978). Indeed, chronic intoxication syndromes including among others visual loss, tropical ataxic neuropathy and Konzo have been discussed more recently in connection with the intake of cyanogens through a monotonous cassava based diet (Brimer and Rosling, 1993; Osuntokun, 1994; Howlett, 1994; Tylleskar, 1994). Konzo, characterized by abrupt onset of a varying degree of symmetrical, isolated, and permanent but non-progressive spastic paraparesis (Tylleskär et al. , 1993) still is around (Nhassico et al. 2008; Chabwine et al. 2011; Cliff et al, 2011).

The cultivation of bitter cassava varieties and thereby roots high in their content of cyanogenic glycosides is considerable in a world perspective (Nyirenda et al. 2011). Such roots cannot legally be traded internationally as food but must be processed locally to reduce the content to safe levels. However, what are the safe levels?

Based on the data available for the toxicity of cyanide and cyanogenic glycosides JEFCA (Joint FAO/WHO Expert Committee on Food and Additives) (1993) tried to make a risk assessment and propose a regulation concerning the safe level for the intake of cyanogenic glycosides for human. However, the committee only could conclude that because of lack of quantitative toxicological and epidemiological information, a safe level of intake of cyanogenic glycosides could not be estimated. Therefore, the committee instead concluded that a level of up to 10 mg kg^{-1} (product) hydrogen cyanide in the Codex Standard for Cassava Flour (approved by Codex Alimentarius Commission, 1991) is not associated with acute toxicity (Speijers, 1993). In 2003, the Codex Alimentarius Commission as earlier mentioned launched a "Codex Standard for Sweet Cassava". According to this, sweet cassava varieties are those that contain less than 50 mg kg^{-1} hydrogen cyanide (fresh weight basis) (CAC, 2003). Later bitter fresh cassava roots were defined not only by a CAC standard (CAC, 2010) but also in a standard from the East African Community (EAC). In accordance with the CAC standard the latter defines bitter cassava roots as roots that release more than 50 mg HCN kg^{-1} fresh weight. However, this standard adds: "when tested in accordance to CD/U/07/2009, sweet cassava roots shall release no more than 200 mg/kg of hydrogen cyanide (fresh weight basis)" (EAC, 2010). Turning to national food agencies, the Food Standards Australia New Zeeland (FSANZ) due to increased import conducted a safety evaluation of cyanogenic constituents in cassava and cassava by-products. It is stated that (i) "by adequate processing (peeling, slicing and cooking) both the cyanogenic glycosides and the hydrocyanic acid can be removed prior to consumption" (FSANZ, 2004), (ii) "while the current users have adequate knowledge regarding the risks associated with the consumption of cassava more widespread use in the community would increase the public health risks" (FSANZ, 2004) and (iii) "a maximum level of 10 mg HCN kg^{-1} is considered to be required to confidently protect public health and safety when looking at ready-made cassava chips" (FSANZ, 2009). Furthermore, the Indonesian Ministry of Industry and Trade set the food safety level for "cyanide" content at 40 mg HCN kg^{-1} fresh weight (Hidayat et al. 2000). The European Union's regulation of "HCN" in manioc by-products intended for animal feed is 100 mg kg^{-1} relative to a feedingstuff with a moisture content of 12% (EFSA, 2007).

Overall, not all countries neither in the developed world nor in the developing world have set official standards for the acceptable maximum content of cyanogenic glycosides and other cyanogens in cassava and cassava products for food nor feed use. Indeed, more scientific investigations as followed by risk assessments are needed in this area of chemical food safety. This because, several recent studies conducted in the developing as well as in the developed world have shown that consumers are aware of the availability of fresh cassava roots as a "new" raw material for the preparation of food (Koling-hansen and Brimer, 2010; Nyirenda et al. 2011). Because of the presence of the cyanogens that may exceed the safe limits different processing and tasting methods have been developed by the populations used to consume cassava to ensure the safety of the products (Chiwona-Karltun et al. 2004; Oluwole et al. 2007; Essono et al. 2008; Nyirenda et al. 2011; Gnonlonfin et al. 2011). Obviously, it is essential that such knowledge is spread, and that information about the content of cyanogens in specific roots and leaves as well as in cassava based products is available.

Fate of Cyanogenic Glycosides during Processing of Cassava

Cassava is one of very few food plants for which a number of toxic varieties deliberately are grown. With toxic is in this context meant, that the harvested material has to be processed; if not for other reasons at least in order to reduce the content of the cyanogenic glycosides present. A large number of scientific studies are available concerning the fate of cyanogenic glycosides during different types of processing.

Linamarin (≥ 90%) and lotaustralin (≤10%) are the constituents of importance in the edible root parenchyma and in the leaves. In both compounds the sugar part is β-D-glucose; the compounds are cyanogenic glucosides. As such they are not very toxic, since they (to a great extent) are absorbed in the gastrointestinal tract and excreted unchanged in the urine. However, during chewing and processing the glucosides normally will be degraded in a reaction cascade starting with hydrolysis of the glycosidic bond followed by cleavage of the cyanohydrin formed; one of the end products being the toxic HCN (hydrogen cyanide). Hydrolysis is catalyzed by a specific β-glucosidase (called linamarase) present in the plant tissue separated from the glucosides. Upon tissue damage the compartmentation is broken and the hydrolysis starts. The cyanohydrins formed are stable at acidic pH while they are spontaneously cleaved under alkaline conditions (Brimer, 2010; OECD, 2009). Upon ingestion they are sources of HCN, since they either will be degraded in the gastrointestinal tract or absorbed and then degraded to release HCN.

Some degree of glucoside hydrolysis cannot be avoided during processing of cassava tubers or leaves. Also hydrolysis may occur during the process of eating the food; in case the linamarase has not been denaturated. Investigations also show that hydrolysis in the gastrointestinal tract cannot be totally ruled out (Carlson et al. 1999). Therefore, processing of toxic cassava varieties must, in addition to steps to improve shelf life, include steps that ensure proper removal of the cyanogens (i.e. the glucosides, the cyanohydrins and the HCN) present. The importance of keeping a focus on this was recently illustrated by the finding of relatively high concentrations of total cyanogens in two investigations of traded products in Zambia/Malawi (Nyirenda et al. 2011) and in Australia (Miles et al. 2011), respectively. The latter analysed imported ready-to-eat chips.

Looking into the published studies about processing inter-relationships for cassava roots it becomes clear (Westby, 2002) that fresh peeled (maybe slightly subdivided) roots may be subjected to one of the following documented unit processes: drying, soaking (retting, steeping), cooking, frying, size reduction (grating), or fermentation (dry in heaps or wet - by extended soaking or after grating). Most cassava root based food products, whether being dishes in them self or being products, such as e.g. flour, will have undergone more than one unit process; i.e. being a result of a "unit process chain". Also, some products will be a result of additional unit processes such as: pressing/dewatering (of grated root material), steaming (e.g. of size reduced and pounded root material), roasting or smoking.

UNIT PROCESSES AND THEIR IMPACT ON THE COMPOSITION OF CASSAVA ROOT BASED FOOD PRODUCTS

Drying

Drying of fresh peeled cassava roots - or more often "chips" (pieces) - gives a decrease in the total content of cyanogens. Thin chips dries faster than thicker chips, the latter showing a higher decrease in cyanogenic potential. This phenomenon is caused by the faster drying of the thin ones, which reduces the contact time between the linamarase and the glucosides. A number of drying studies are reviewed by Padmaja (1995).

A recent study showed a strong increase in the content of the coumarin compound scopoletin (7-hydroxy-6-methoxychromen-2-one) after peeling, cutting into slices of a thickness of 2 cm (chipping) and drying for 6 days. The mean content in fresh roots from four varieties varied between 4.1 to 11.1 mg kg^{-1}, while the content after the 6 days rose to 242 mg kg^{-1} for the cultivar showing the highest increase (Gnonlonfin et al., 2011).

Soaking

During the soaking of roots in water the texture undergoes changes and the roots are rendered soft. Attempts have been made to quantify softening and use the result as an index to monitor the extent of fermentation. Softening of tissues can be attributed to action of pectinolytic and cellulytic enzymes produced by microorganisms. Softening can be helpful in subsequent size reduction operations such as crushing and grinding. Studies have been made in plain water (allowing the influence of the natural microbial flora) and after the addition of microbial inoculums. The influence of soaking on the content of cyanogens are overviewed by Padmaja (1995), who quotes a study where Bourdoux et al. found that soaking roots in water for 1 day decreased the total cyanogenic potential from 108 to 60 mg kg^{-1}, while a 5 day treatment resulted in a final content of 3 mg kg^{-1} (Padmaja, 1995).

Cooking (Baking, Frying and Steaming)

Cooking of cassava root material cut into pieces, as documented from eastern Nigeria and as part of food preparation in Sri Lanka and South America reduces the content of total cyanogens. The reduction depends on parameters, such as boiling time, volume of water as related to volume of cassava material and the tuber piece size. Most often the reduction will not be higher than around 50% (maximum 70%) (Padmaja, 1995). An investigation using three different varieties of cassava showed that cooking in general is more effective in reducing the total cyanogenic potential than baking, frying and steaming (Nambisan and Sundaresan, 1985).

Cooking can extract/degrade the content of carotenoids and promote isomerization in β-carotene; increasing the concentration of Z isomers with low pro-vitamin A activity. Using seven different cultivars of yellow sweet cassava with a mean total β-carotene content varying between 1.99 and 8.11 µg/g (f.w.) three different 30 minutes cooking processes were tested. The percentual retention of β-carotene varied both with cultivar and cooking method. However, most cultivars and methods showed retention in the range from around 65 to 95% (Carvalho et al. 2012).

Size Reduction (Grating)

If not processed into chips, most processing of cassava roots initially involves a size reduction of peeled fresh or soaked roots by grating, resulting in wet pulp. Studies by Hahn performed in West Africa showed pulp made from fresh roots to be processed directly by molding it into balls of a diameter around 3-5 cm and put into boiling water (referred through Gnonlonfin et al. 2012a).

Wet pulp from fresh roots may also be fermented as is the case in the production of gari (Onabolu et al. 2002a,b) and agbelima (Amoa-Awua et al. 1996), respectively, the latter being added an inoculum. Wet pulp produced by soaking peeled roots for about three days or unpeeled for up to five day, squeezing the roots to remove water and then pounding of the material have been described from different regions of the Democratic Republic of Congo. After some kind of drying the dried material is again pounded to flour (Banea et al. 1992).

The literature is limited concerning investigations measuring the concentrations of linamarin, other cyanogens and free HCN, respectively. It can be concluded, however, that the finer the grating the more cell disruption, which immediately initiates the degradation of linamarin to the corresponding cyanohydrin.

Fermentation (Dry in Heaps or Wet - by Extended Soaking or after Grating)

Heap fermentation of roots takes place in Uganda, Congo and Mozambique (Tivana et al. 2007). In a Mozambique study the average cyanogenic content of the peeled fresh root material was around 260 mg HCN equivalents kg^{-1}. Flour made from material heap fermented for four days as followed by sun drying had a total cyanogenic potential of 17 mg HCN equivalents kg^{-1}, which can be compared to flour made from directly sun dried material, showing a content of 158 mg HCN equivalents kg^{-1}(Tivana et al. 2007).

Changes in cyanogens during grating and natural lactic fermentation of cassava roots at 30°C to form gari were studied by Vasconcelos and co-workers. Within three hours of grating more than 95% of the initial linamarin content was hydrolyzed. Significant concentrations of cyanohydrin (and free HCN) were left in the mash. Under the acidic conditions formed as a result of the fermentation the cyanohydrins were rather stable. After 100 h the total cyanogenic potential thus only had decreased from around 170 to approximately 75 mg HCN eq. kg^{-1} (d.w.) (Vasconcelos et al. as cited through Westby, 2002). The further changes in the content of cyanohydrins in a cassava mash and in the gari (after roasting) were studied in a model production of gari in Nigeria. Continuous dewatering of the mash by pressure during the fermentation resulted in higher contents than did dewatering after fermentation (Onabolu et al. 2002a,b).

The changes in total protein percentage, mineral content, vitamin content and amino acid composition during fermentation of cassava roots have been reviewed by Moorthy and Mathew (1998). Losses of approximately 50% or more is seen for minerals such as Ca, Mg, K, Fe, P, Na and Mn. Loos of zink seems to be higher while Cu was retained. Protein content may be slightly enhanced as a result of fermentation, however, also considerable losses (up to around 50%) have been reported by Moorthy and Mathew (1998). A large reduction in ascorbic acid from around 80 and to around 7.5 mg kg^{-1} (f.w.) during gari fermentation was attributed to leaching out of the ascorbic acid into the steep water (Moorthy and Mathew, 1998).

CASSAVA AND MYCOTOXINS: AFLATOXINS AND OTHER IMPORTANT MYCOTOXIN

The presence of low levels of aflatoxins has been reported in cassava chips from Tanzania and the Republic of Congo with the mean ranging from 0.1 to 1.3 μg kg^{-1} and in other cassava products ranging from 0.3 to 4.4 μg kg^{-1} (Manjula et al. 2009). Among four samples of cassava flour from Benin two were found to contain no aflatoxin B_1 while the two others were shown to be contaminated at a low level of up to 9 μg kg^{-1} (Ediage et al. 2011). In contrast, other works conducted in Tanzania (Muzanila et al. 2000), Nigeria (Jimoh and Kolapo, 2008) and investigations from Benin (Gnonlonfin et al. 2008; Gnonlonfin et al. 2012b) have revealed no aflatoxin contamination in cassava chips, the limit of detection in the later study being 0.1 μg kg^{-1}. On the other hand other mycotoxin such as neosolaniol, T-2 toxin (*Fusarium* sp), patulin, cyclopiazonic acid, penicillic acid and diacetoxyscirpenol were identified in cassava chips flour from Uganda and Ivory coast Westby et al. (1994). A later study found and quantified a number of these compounds, i.e. sterigmatocystin (0.17-1.67 mg kg^{-1}), patulin (0.55-0.85 mg kg^{-1}), cyclopiazonic acid (0.08-0.72 mg kg^{-1}), penicillic acid (0.06-0.23 mg kg^{-1}) and tenuazonic acid (0.02-0.34 mg kg^{-1}) in Ghanaian cassava products such as dried cassava chips so-called "*Kokonte*" (Wareing et al. 2001). The occurrence of diacetoxyscirpenol at a level up to 9 μg kg^{-1} in cassava flour was confirmed recently in a study which also found indeed very low levels of zearalenone (up to 12 μg kg^{-1}) and fumonisin B_1 (up to 219 μg kg^{-1}) (Ediage et al. 2011).

Recently, several studies aimed at establishing prevalence of key mycotoxins have been initiated, although only a few have been carried out in Tanzania, Nigeria, Cameroon, Congo Brazzaville, Benin, and all have focused on aflatoxin and fumonisins. Exposure to large doses of individual or mixed mycotoxins may cause acute toxicity with fatal effect.

Aflatoxins

The acute toxicity of one of the most toxic mycotoxins i.e. aflatoxin B1 has been demonstrated to vary very much between animal species. Thus, the acute lethal effect is reported to show an oral LD_{50} value of around 5 mg kg^{-1} body weight (b.w.) for the rat but between 0.4 to 0.5 mg kg-1 b.w. for the rabbit and the cat (Wagacha and Muthomi, 2008; ChemIDPlus Lite, 2010). Severe acute liver injury with high morbidity and mortality has been associated with high dose of aflatoxins exposure (Chao et al. 1991). Estimated mixed total doses of aflatoxins (B1, B2, G1, G2) ingested with food of around 6000 mg have been reported to cause fatal acute toxicity in adult human beings. Some of the effects related to the acute toxicity include reduced liver function, disruption of blood clotting mechanism, icterus (jaundice) and a decrease in essential serum proteins that are synthesized by the liver (Wagacha and Muthomi, 2008). Other symptoms of acute to subacute aflatoxicosis are oedema of the lower extremities, abdominal pain, and vomiting. No animal species is resistant to the acute toxic effects of aflatoxins, however, species respond differently in their susceptibility to acute and chronic toxicity of mycotoxins. Exposure to small doses (a total of 2-6 mg) distributed over a prolonged period can lead to cancer as already mentioned (Groopman and Kensler, 1999; Wagacha and Muthomi, 2008). In general the EU is recognized as the region with most strict regulation concerning mycotoxins in food. The maximum accepted levels of different mycotoxins in food in the EU is laid down in Commission Regulation (EC) No. 1881/2006, as amended (EU, 2006), currently for aflatoxins (B_1, total (B_1+B_2+G_1+G_2) and M_1), ochratoxin A, patulin, deoxynivalenol, zearalenone and fumonisins (B_1+B_2). Accepted aflatoxin levels for commodities to be subjected to sorting ranges (depending on commodity) from 5-12 μg kg^{-1} for AFB_1 (10-15 for the sum) while the ranges of levels for commodities intended for direct human consumption (except foods for infants and young children) are AFB_1 (2-8) and total (4-10) μg kg^{-1}. In this perspective the levels found for aflatoxin contamination in cassava are absolutely reasonable.

Sterigmatocystin

High concentrations of sterigmatocystin (STC) was found in the cassava chips from Ghana (concentration range 0.17-1.67 mg kg^{-1}) using an HPLC method (Wareing et al. 2001).

In contrast to the aflatoxins, STC has a low acute toxicity which among others may be due to its poor absorption rate from the digestion tracts (Terao, 1983). A very early study has shown that the oral LD_{50} in rats was 120-166 mg kg^{-1}, but part of this toxicity may be attributable to dimethylformamide used as solvent (Purchase and Van Der Watt, 1969).

In conclusion it seems that cassava and cassava products in general are low in the content of aflatoxins, while the structurally related STC may be present in significant concentration. However, the later needs to be further investigated since based on only one study.

Patulin

Also patulin was found in cassava chips products from Ghana (Wareing et al. 2001)-. Patulin is relatively heat stable and is not destroyed by pasteurization (Bara, 2009). It does not appear to survive fermentation processes and is not usually found in alcoholic drinks, such as cider, but the toxicity of its breakdown products is uncertain (Pfohl-Leszkowicz, 1999). Most of the information on the toxicity of patulin is derived from animal studies and there is little or no epidemiological data on acute or chronic toxicity in humans. Of more concern from a food safety point of view are longer term chronic effects. It was early suggested that patulin could be a carcinogen at low levels in the diet, but the International Agency for Research on Cancer (IARC) has reviewed the available data and concluded that there is no convincing evidence of carcinogenicity in animals or in humans, other than at extremely high doses (IARC, 1986). Data from feeding experiments (mice) have been used to derive a no observed effect level (NOEL) of 43 μg kg^{-1} b.w. per day and a provisional maximum tolerable daily intake (PMTDI) for humans of 0.4 μg kg^{-1} b.w. (WHO, 1998). This is well above the maximum daily intake levels estimated by WHO for adults (0.1 μg kg^{-1} b.w.) and children (0.2 μg kg^{-1} b.w.) (WHO, 1998). The concentrations of patulin (0.55 – 0.85 mg kg^{-1}) found in cassava may seem high, however, it should be remembered that the dry matter content of cassava chips is much higher than in the EU regulated products. On the other hand cassava may constitute a higher proportion of the daily diet. Further investigations clearly are needed.

Cyclopiazonic Acid

Cyclopiazonic acid (CPA) is an indole tetramic acid mycotoxin produced by the ubiquitous genera of molds *Aspergillus* and *Penicillium*. Besides colonizing various grains and seeds (Abbas et al. 2008; Njobeh et al. 2009), these fungi can grow on many food substrates including cassava products (Gnonlonfin et al. 2008, 2012b). CPA can contaminate numbers of agricultural commodities, animal feeds and food sources. Despite the wide presence of CPA, few incidents of animal mycotoxicosis and no confirmed incident of human poisoning have been attributed to CPA. It is not a very potent acute toxin, its oral LD_{50} in rats for instance is in the range of 30-70 mg kg^{-1} (Nishie et al. 1987; Antony et al. 2003). It has been speculated that CPA may be associated with “kodo poisoning”, a toxic reaction characterized by nausea, vomiting, depression, and unconsciousness after consumption of CPA-contaminated kodo millet in North India (Roa and Husain, 1985). Based on a NOEL of 1 mg kg^{-1} day^{-1} and taking into consideration species variability, for humans approximately 10 μg kg^{-1} day^{-1} or 700 μg day^{-1} was suggested as the maximum acceptable daily intake (ADI) for CPA by Burdock and Flamm (2000). The finding of CPA concentrations in the range of 0.08 – 0.72 mg kg^{-1} in cassava chips seems to give no reason for alarm.

Penicillic Acid and Tenuazonic Acid

Penicillic acid (PA) is a carcinogenic mycotoxin that was early isolated from agricultural products such as corn, poultry feed, dried beans, tobacco and cassava products (Kurtzman and Ciegler, 1970; Pero et al. 1972; Bacon et al. 1973; Thorpe and Johnson, 1974; Wareing et al. 2001). Human carcinogenicity data, toxicity case reports or epidemiological studies are not available. No authorities have regulated PA in food. The finding of PA concentrations in the range of 0.55 – 0.85 mg kg^{-1} in cassava chips seem not to be harmful.

Tenuazonic acid (TA) was first isolated from sorghum grains and it has been associated with a human haematological disorder known as “Onyalai” (Steyn and Rabie, 1976; Bottalico and Logriego, 1998). Toxicity data for tenuazonic acid relevant for hazard characterization are not available. No authorities have regulated TA in food.

Data available on the toxicity of PA and TA do not allow any risk assessment and in general the overall findings on mycotoxins and cassava reported here are thus rather spotty and isolated, calling for more investigations focusing on both cassava and different cassava products. Summarizing the results of investigations on the occurrence of mycotoxin in cassava we find that potential consumer exposure seems to be low.

Diacetoxyscirpenol, Zearalenone and Fumonisins

Trichothecenes are classified into four groups (type A, B, C, and D) based on their chemical structures as well as their producer fungi. Among these, diacetoxyscirpenol (DAS) and T-2 toxin in type A, and deoxynivalenol (DON) and nivalenol (NIV) in type B, are known to be the most naturally encountered and the most potent trichothecenes (Jun et al., 2007). While DON is regulated with accepted maximum levels ranging from 500 to 1750 μg kg^{-1}(except for products for children) in the EU (EU, 2006), DAS is not regulated here or elsewhere. In this connection the reported findings of DAS in cassava products seems negligible. Zearalenone was only found in one investigation of cassava so far in concentrations up to 12 μg kg^{-1} (Ediage et al. 2011). The lowest maximum concentration in the EU regulation being 20 μg kg^{-1} this seems to give rise to no concern. Fumonisin B_1 was likewise found in concentrations up to 219 μg kg^{-1} by Ediage et al. (2011). The range of accepted maximum concentration for general commodities in the EU regulation being 800 μg kg^{-1} , again this is of no concern (EU, 2006).

CASSAVA AND COUMARINS

Coumarin and its derivatives are found in the plant kingdom both mono- and dicotyledons. Coumarin itself is produced by plants as a defence chemical to discourage predation. Scopoletin (6-methoxy-7-hydroxycoumarin) is a coumarin compound that has been isolated from cassava cultivars from Colombia (Bushmann et al. 2000) and more recently from Benin (Gnonlonfin et al. 2011b). Investigations have shown that scopoletin inhibited aflatoxin biosynthesis even at lover concentrations of 0.024 mM (Gnonlonfin et al. 2011a).

Humans metabolize coumarin to 7-hydroxycoumarin, a compound of lower toxicity. The German Federal Institute for Risk Assessment (GFIRA) has established a tolerable daily intake of 0.1 mg coumarin per kg b.w., but also stte that if this level is exceeded for a short time only, there is no threat to health. For instance, a person weighing 60 kg would have a TDI of approximately 6.0 mg of coumarin. Compared to concentration of scopoletin found (4.1 – 242.5 mg kg^{-1}) in newly produced dry cassava chips decreasing rapidly to 0.7 mg kg^{-1} after 3 months of storage (Gnonlonfin et al. 2012), and assuming a lower toxicity in general for the different hydroxylated coumarins, than for coumarin itself, consumption of cassava and cassava products seems to be safe at least with regards to acute toxicity of hydroxycoumarins. This in spite of the fact that earlier report suggested cassava toxicity may be partly due to coumarins (Dall'Acqua et al. 1978; Downum, 1992; Bourgaud et al. 2006).

CONCLUSION

Cassava root tuber production and intake of products thereof is increasing worldwide with the introduction and adoption of new varieties/cultivars in areas where this crop was not known, and with increasing trade internationally However, food safety requirements differ from country to country and can constitute an important barrier for international trade in food products, as their main objective is to protect public health (Broberg, 2009).

Overall, this short excursus indicates that cassava and cassava products when handled correctly in general are safe food. However, good agricultural and processing practices have to be implemented to ensure the safety of the raw or final products. Furthermore, on the basis of the review presented there is still a need to set regulatory limits concerning the acceptable concentrations of the cyanogenic compounds at national, regional and international level.

ACKNOWLEDGMENTS

The authors are grateful to Danish International Development Assistance (DANIDA) for financial support. The authors gratefully acknowledge all mycotoxicologists, chemists and mycologists for their excellent work that was the basis for this critical review.

REFERENCES

Abbas, H. K., Accinelli, C., Zablotowicz, R. M., Abel, C. A., Bruns, H. A., Dong, Y. & Shier, W. T. (2008). Dynamics of mycotoxin and *Aspergillus flavus* levels in aging Bt and non-Bt corn residues under Mississippi no-till conditions. *J. Agric. Food Chem. 56*, 7578-7585.

AbdAlla, E. A., Metwally, M. M., Mehriz, A. M. & Abu Sree, Y. H. (1996). Sterigmatocystin, incidence, fate and production by *Aspergillus versicolor* in Ras cheese. *Nahrung*, *40*, 310-313.

Adebajo, L. O. (1993). Survey of aflatoxins and ochratoxin A in stored tubers of *Cyperus esculentus* L. *Mycopathologia*, *124*, 41-46.

Akano, D. A. & Atanda, O. O. 1990. The present level of aflatoxins in Nigerian groundnut cake ('kulikuli'). *Lett. Appl. Microbiol. 10*, 187-189.

Amoa-Awua, W. K. A., Appoh, F. E. & Jakobsen, M. (1996). Lactic acid fermentation of cassava dough into agbelima. *Int. J. Food Microbiol. 31*, 87-98.

Antony, M., Shukla, Y. & Janardhanan, K. K. (2003). Potential risk of acute hepatotoxicity of kodo poisoning due to exposure to cyclopiazonic acid. *J. Ethnopharmacol. 87*, 211-214.

Atalla, M. M., Hassanein, N. M., El-Beih, A. A. & Youssef, Y. A. (2003). Mycotoxin production in wheat grains by different *Aspergilli* in relation to different relative humidities and storage periods. *Nahrung*, *47*, 6-10.

Azziz-Baumgartner, E., Lindblade, K., Gieseker, K., Rogers, S., Kieszak, S., Njapau, H., Schleicher, R., McCoy, L., Misore, A., DeCock, K., Rubin, C. & Slutsker, L. (2005). Aflatoxin investive group. Case-control study of an acute aflatoxicosis outbreak, Kenya, 2004. *Environ. Health Pers. 113*, 1779-1783.

Bacon, C. W., Sweeney, J. G., Robbins, J. D. & Burdick, C. (1973). Production of penicillic acid and ochratoxin A on poultry feed by *Aspergillus ochraceus*, temperature and moisture requirements. *Appl. Microbiol. 26*, 155-160.

Bando, M., Hasegawa, M., Tsuboi, Y., Miyake, Y., Shiina, M., Ito, M., Handa, H., Nagai, K. & Kataoka, T. (2003). The mycotoxin penicillic acid inhibits fas ligand-induced apoptosis by blocking self-processing of caspase-8 in death-inducing signaling complex. *The J. Biol. Chem. 278*(8), 5786-5793.

Banea, M., Poulter, N. H. & Rosling, H. (1992). Shortcuts in cassava processing and risk of dietary cyanide exposure in Zaire. *Food Nutr Bull 14*, 137-143.

Bankole, S. A. & Mabekoje, O. O. (2004). Mycoflora and occurrence of aflatoxin B1 in dried yam chips from markets in Ogun and Oyo States, Nigeria. *Mycopathologia*, *157*, 111-115.

Bankole, S. A., Ogunsanwo, B. M. & Eseigbe, D. A. (2005). Aflatoxins in Nigerian dry-roasted groundnuts. *Food Chem. 89*, 503-506.

Bankole, S. A., Ogunsanwo, B. M. & Eseigbe, D. A. (2005). Aflatoxins in Nigerian dry-roasted groundnuts. *Food Chem. 89*, 503-506.

Bara, C. (2009). Wine contamination with patulin. *Natural Resources and Sustainable Development*, 199-202.

Barug, D., Van Egmond, H., Lpez-Garcia, R., Van Osenbruggen, T., Visconti, A. (2003). Meeting the mycotoxins menace. Wageningen Academic Publishers, The Netherlands.

Battilani, P., Costa, L. G., Dossena, A., Gullino, M. L., Merchelli, R., Galaverna, G., Pietri, A., Dall'Asta, C., Giorni, P., Spadaro, D., Gualla, A. (2009). *Scientific information on mycotoxins and natural plant toxicants*. CFP/EFSA/CONTAM/2008/01. Scientific Technical Report 379-467.

Beasley, V. (1999). Mycotoxins that affect the liver. In, Beasley V. (ed.) Veterinary Toxicology. *International Veterinary Information Service*, New York, USA.

Beeching, J. R., Han, Y., Gomez-Vasquez, R., Day, R. C., Cooper, R. M. (1998). Wound and defense responses in cassava as related to post-harvest physiological deterioration. *Recent Adv. Phytochem. 32*, 231-248.

Bhat, R. V. (1991). Aflatoxins, Successes and failures of three decades of research. *ACIAR proceedings*, *36*, 77-82.

Bokanga, M. (1994a). Distribution of cyanogenic potential in cassava germplasm. *Acta Hortic.* (Wageningen) *375*, 117-123.

Bokanga, M. (1994b). Processing of cassava leaves for human consumption. *Acta Hortic.* (Wageningen) *375*, 203-207.

Bottalico, A. & Logrieco, A. (1998). Toxigenic *Alternaria* species of economic importance. In Sinha KK, Bhatnagar D. (Eds). *Mycotoxins in agriculture and food safety*. Marcel Dekker, Inc, New York, USA, 65-108.

Bourgaud, F., Hehn, A., Larbat, R., Doerper, S., Gontier, E., Kellner, S. & Matern, U. (2006). Biosynthesis of coumarins in plants, a major pathway still to be unraveled for cytochrome P450 enzymes. *Phytochem. Rev. 5*, 293-308.

Brimer, L. (2001). Chemical hazards and their control, endogenous compounds. In, Adams MR, Nout MJR. (Eds), Fermentation and food safety. Aspen Publishing, *Gaithersburg*, 71-98.

Brimer, L. (2010). Cyanogenic Glycosides. In, Bernhoft A (ED), the Proceedings of the Symposium Bioactive compounds in plants – benefits and risks for man and animals. Norwegian Academy of Sciences and Letter. The Norwegian Academy of Sciences and Letters/Novus, *Oslo*, 125-143.

Brimer, L. & Rosling, H. A. (1993). Microdiffusion method with solid state detection for determination of cyanogenic glycosides from cassava in human urine. *Food Chem. Toxicol. 31*, 599-603.

Broberg, M. (2009). *European food safety regulation and the developing countries, regulatory problems and possibilities*, 44. Working paper 2009:09, Danish Institute for International Studies, DIIS, Denmark.

Bryden, W. L. (2007). Mycotoxins in the food chain, human health implications. *Asia Pac. J. Clin. Nutr. 16*(Suppl 1), 95-101.

Burdock, G. A. & Flamm, W. G. (2000). Review article, Safety assessment of the mycotoxin cyclopiazonic acid. *Int. J. Toxicol. 19*, 195-218.

Burns, A. E., Bradbury, J. H., Cavagnaro, T. R. & Gleadow, R. M. (2012). Total cyanide content of cassava food products in Australia. *J. Food Comp. Anal. 25*, 79-82.

Buschmann, H., Rodriguez, M. X., Tohme, J. & Beeching, J. R. (2000). Accumulation of hydroxycoumarins during post-harvest deterioration of tuberous roots of cassava (Manihot esculenta Crantz). *Ann. Bot. 86*, 1153-1160.

CAC (Codex Alimentarius Commission) (2003). Codex standard for sweet cassava. Codex Stan 238-2003, Amended in 2005, 4.

CAC (Codex Alimentarius Commission) 2010. Codex standard for bitter cassava. *Codex Stan* 300-2010, 4.

Carlsson, L., Mlingi, N., Juma, A., Ronquist, G. & Rosling, H. (1999). Metabolic fates in humans of linamarin in cassava flour ingested as stiff porridge. *Food Chem. Toxicol. 37*, 307-312.

Carvalho, L. M. J., Oliveira, A. R. G., Godoy, R. L. O., Pacheco, S., Nutti, M. R., de Carvalho, J. L. V., Pereira, E. J. & Fukuda, W. G. (2012). Retention of total carotenoid and β-carotene in yellow sweet cassava (Manihot exculenta Crantz) after domestic cooking. *Food Nutr. Res. 56* (15788 – DOI, 10.3402/fnr.56i0.15788)

CEC (Commission of the European Communities). (1998). EC directive 98/1525.16.07.1998. *Off. J. EC. L 201*, 43.

Chabwine, J. N., Masheka, C., Balol'ebwami, Z., Maheshe, B., Balegamire, S., Rutega, B., Wa Lola, M., Mutendela, K., Bonnet, M. J., Shangalume, O., Balegamire, J. M. & Nemery, B. (2011). Appearance of konzo in South-Kivu, a wartorn area in the Democratic Republic of Congo. *Food Chem. Toxic. 49*, 644-649.

Chan, P. K. & Hayes, A. W. (1982). In vitro metabolism of penicillic acid with mousse-liver homogenate fractions. *Food Chem. Toxic. 20*, 61-65.

Chao, T. C., Maxwell, S. M. & Wong, S. Y. (1991). An outbreak of aflatoxicosis and boric acid poisoning in Malaysia, a clinicopathological study. *J. Pathol. 164*, 225-333.

ChemIDPlus Lite. (2010). United States National Library of Medicine (seen 01.10.2010 at http://chem.sis.nlm.nih.gov/chemidplus/ProxyServlet?objectHandle=Search&actionHand

le=getAll3DMViewFiles&nextPage=jsp%2Fcommon%2FChemFull.jsp%3FcalledFrom %3Dlite&chemid=0001162658&formatType=_3D).

Chiwona-Karltun, L., Brimer, L., Saka, J. D. K., Mhone, A. R., Mkumbira, J., Johansson, L., Bokanga, M., Mahungu, N. M. & Rosling, H. (2004). Bitter taste in cassava roots correlates with cyanogenic glucosides levels. *J. Sci. Food Agric. 84*, 581-590.

Chiwona-Karltun, L., Mkumbira, J., Saka, J., Bovin, M., Mahungu, N. M. & Rosling, H. (1998). The importance of being bitter – a qualitative study on cassava cultivar preference in Malawi. *Ecol. Food Nutr. 37*, 219-245.

Chun, H. S., Ok, H. E., Kim, H. J., Hwang, J. B. & Chung, D. H. (2006). *Risk assessment of aflatoxins in food products consumed in South Korea*. doi, 10.1051/IUFoST:20060458. http://dx.doi.org/10.1051/IUFoST:20060458.

Cliff, J., Muquingue, H., Nhassico, D., Nzwalo, H. & Bradbury, J. H. (2011). Konzo and continuing cyanide intoxication from cassava in Mozambique. *Food and Chemical Toxicology*, *49*, 631-635.

Dall'Acqua, F., Vedaldi, D. & Recher, M. (1978). The photoreaction between furocoumarins and various DNA with different base compositions. *Photochem. Photobiol. 27*, 33-36.

Delage, N., d'Harlingue, A., Colonna Ceccaldi, B. & Bompeix, G. (2003). Occurrence of mycotoxins in fruit juices and wine. *Food Contr. 14*, 225-227.

Dofour, D. L. (1993). The bitter is sweet, a case study of bitter cassava (*Manihot esculenta*) use in Amazonia. In, Hladik C, Hladik A, Linares O, Pagezy H, Semple A, Haldey M. (Eds.), Tropical forests, people and food, Biocultural interactions and applications to development. UNESCO/Parthenon, Paris, pp. 575-588.

Dorner, J. W., Cole, R. J., Lomax, L. G., Gosser, H. S. & Diener, U. L. (1983). Cyclopiazonic acid production by *Aspergillus flavus* and its effects on broiler chickens. *Appl. Environ. Microbiol. 46*, 698-703.

Dorozinsky, A. (1978). Cassava may lead to mental retardation. *Nature*, *272*, 121.

Downum, K. R. (1992). Light-activated plant defence. *New Phytol. 122*, 401-420.

Dunstan, W. R., Henry, T. A. & Auld, S. J. M. (1906). Cyanogenesis in plants V. the occurrence of phaseolutin in cassava (*Manihot aipi* and *Manihot utilissima*). *Proc. R. Soc. Lond. SER. B Biol. SCI. 78*, 152-158.

EAC (East African Community) (2010). *East African Standard*. Bitter cassava – specification and quality (CD/K/106:2010 (ICS 67.080.10). Arusha, EAC.

EC (European Commission). (2006). The commission decision, 2006/504/EC. *Off. J. Eur. Union. L199*, 21-32.

EC (European Commission). (2006). Commission regulation (EC) No 1881/2006. *Off. J. Eur. Union*. L364/24.

Ediage, E. N., Di Mavungu, J. D., Monbaliu, S., Van Peteghem, C., De Saeger, S. (2011). A Validated Multianalyte LC_MS/MS Method for Quantification of 25 Mycotoxins in Cassava Flour, Peanut Cake and Maize Samples. *J. Agric. Food Chem. 59*, 5173-5180.

EFSA (European Food Safety Authority). (2007). Opinion of the scientific panel on contaminants in the food chain on a request from the commission related to cyanogenic compounds as undesirable substances in animal feed. Question no EFSA-Q-2003-064. *EFSA J. 434*, 1-67.

El-Sharkawy, M. A. (2003). Cassava biology and physiology. *Plant Mol. Biol. 53*, 621-641.

El-Tinay, A. H., Bureng, P. L. & Yas, E. A. E. (1984). Hydrocyanic acid levels in fermented cassava. *J. Food Technol. 19*, 197-202.

Essers, A. J. A., Ebong, C., van der Nout, R. M., Grift, M. J. R, Otim-Nape, W. & Rosling, H. (1995). Reducing cassava toxicity by heap fermentation in Uganda. *Int. J. Food Sci. Nutr. 46*, 125-136.

Essono, G., Ayodele, M., Foko, J., Akoa, A., Gockowski, J., Ambang, Z., Bell, J. M. & Bekolo, N. (2008). Farmers' perceptions of practices and constraints in cassava (*Manihot esculenta* Crantz) chips production in rural Cameroon. *Afr. J. Biotechnol. 7*(22), 4172-4180.

EU (European Union) (2006). Commision regulation (EC) No 1881/2006 of 19 December setting maximum levels for certain contaminats in foodstuffs (as amended). Available at, http://eur-lex.europa.eu/LexUriServ/LexUriServ.do?uri=CONSLEG:2006R1881:20100701:EN:PDF

Fandohan, P., Zoumenou, D., Hounhouigan, D. J., Marasas, W. F. O., Wingfield, M. J. & Hell, K. (2005). Fate of aflatoxins and fumonisin during the processing of maize into food products in Benin. *Int. J. Food Microbiol. 98*, 249-259.

FAO (Food and Agriculture Organization) (1996). Food and Agricultural Organization. Worldwide regulations for mycotoxins, 1995. *FAO Food Nutr. Paper. 55*. FAO, UN, Rome, Italy.

FAO (Food and Agriculture Organization) (1997). Joint FAO/WHO Expert Committee on Food Additives, June 17-26, Rome, Italy.

FAO (Food and Agriculture Organization) (2000). *The world cassava economy , Facts, trends and outlook*. Rome, Italy.

FDA (Food and Drugs Administration). (2004). Compliance guidance manual. http://www.cfsan.fda.gov.

Fellinger, A. (2006). *Worldwide mycotoxin regulations and analytical challenges*. World Grain Summit, Foods and Beverages, San Francisco, California, USA. 17-20.

FSANZ (Foof Standards Australia New Zealand). (2004). Cyanogenic glycosides in cassava and bamboo shoots. *A human health risk assessment*. Technical report series No. 28, 24.

FSANZ (Foof Standards Australia New Zealand). (2009). Proposal P1002. Hydrocyanic acid in ready-to-eat cassava chips. First review report, 24.

GFIRA (The German Federal Institute for Risk Assessment) 2006. Frequently asked questions about coumarin in cinnamon and other food. http://www.bfr.bund.de/cm/279/frequently_asked_questions_about_coumarin_in_cinnamon_and_other_foods. (pdf).

Gnonlonfin, G. J. B., Adjovi, Y., Gbaguidi, F., Gbenou, J., Moudachirou, M., Brimer, L. & Sanni, A. (2011a). Scopoletin in cassava products as an inhibitor of aflatoxin production. *Journal of Food Safety 31*, 553-558.

Gnonlonfin, G. J. B., Gbaguidi, F., Gbenou, J. D., Sanni, A. & Brimer, L. (2011b). Changes in scopoletin concentration in cassava chips from four varieties during storage. *J. Sci. Food Agric. 91*, 2344-2347..

Gnonlonfin, G. J. B., Hell, K., Fandohan, P. & Siame, A. B. (2008). Mycoflora and occurrence of aflatoxins and fumonisin B1 in cassava and yam chips from Benin, West Africa. *Int. J. Food Microbiol. 122*, 140-147.

Gnonlonfin, B. G., Sanni, A. & Brimer, L. (2012a). Preservation of cassava (Manihot esculenta Crantz): a major crop to nourish people worldwide. In, Bhat R, Alias AK, Paliyath G. (Eds), Progress in Food Preservation. Wiley-Blackwell, Hoboken, 331-342.

Gnonlonfin, G. J. B., Adjovi, Y., Katerere, D. R., Shephard Gordon, S., Sanni, A. & Brimer, L. (2012b). Mycoflora and absence of aflatoxin in commercialized cassava chips in Benin, *West Africa. Food Contr. 23*, 333-337.

Groopman, D. J. & Kensler, T. W. (1999). The light at the end of the tunnel for chemical-specific biomarkers, daylight or headlight? *Carcinogenesis*, *20*, 1-11.

Hell, K., Cardwell, K. F., Setamou, M. & Schulthess, F. (2000). Influence of insect infestation on aflatoxin contamination of stored maize in four agroecological regions in Benin. *Afr. Entomol. 8*, 169-177.

Hell, K., Gnonlonfin, G. J. B., Kodjogbe, G., Lamboni, Y. & Abdourhamane, I. K. (2009). Mycoflora and occurrence of aflatoxin in dried vegetables in Benin, Mali and Togo, West Africa. *Int. J. Food Microbiol. 135*, 99-104.

Henry, S., Bosch, F. X., Troxell, T. C. & Bolber, P. M. (1999). Reducing liver cancer-global control of aflatoxin. *Sci. 286*, 2453-2454.

Herbert, V. C. (1988). In modern nutrition in health and disease, 7th Edition. M.E. Shils, U.R. Young (eds), *Lea and Febiger*, *471*.

Hidayat, A., Zuaraida, I., Hanarida, I. & Damardjati, D. (2000). Cyanogenic content of cassava root of 179 cultivars grown in Indonesia. *J. Food Comp. Analysis*, *13*, 71–82.

Houssou, P., Ahohuendo, B. C., Fandohan, P., Kpodo, K., Hounhouigan, D. J. & Jakobsen, M. (2009). Natural infection of cowpea (*Vigna unguiculata* (L.) Walp.) by toxigenic fungi and mycotoxin contamination in Benin, West Africa. *J. Stored Prod. Res. 45*, 40-45.

Howlett, W. P. (1994). Konzo, a new human disease entity. *Acta Hortic.* (Wageningen) *375*, 325-329.

Hsieh, D. & Atkinson, D. N. (1991). Bisfuranoid mycotoxins, their genotoxicity and carcinogenicity. *Adv. Exp. Med. Biol. 283*, 525-532.

IARC (International Agency of Research on Cancer). (1986). Some naturally occurring and synthetic food components, furocoumarins and ultraviolet radiation. In Monographs on the evaluation of the carcinogenic risk of chemicals to humans; IARC, Lyon, France, 83-98.

IARC (International Agency of Research on Cancer). (1993). Some naturally occurring substances, Food items and constituents, heterocyclic aromatic amines and mycotoxins. Aflatoxins. *WHO IARC Monographs on the evaluation of carcinogenic riks to humans*, Vol. *56*, 245-395. Lyon.

JEFCA (Joint FAO/WHO Expert Committee on Food Additives). (1993). Toxicological evaluation of certain food additives and naturally occurring toxicants (The 39th Meeting of JEFCA). *WHO Food Addi. Ser.* no. 30.

Jiang, Y., Jolly, P. E., Ellis, W. O., Wang, J. S., Phillips, T. D. & Williams, J. H. (2005). Aflatoxin B1 albumin adduct levels and cellular immune status in Ghanaians. *Int. Immunol. 17*, 807-814.

Jiang, Y., Jolly, P. E., Preko, P., Wang, J. S., Ellis, W. O., Phillips, T. D. & Williams, J. H. (2008). Aflatoxin-related immune dysfunction in health and in human immunodeficiency virus disease. *Clin. Dev. Immunol.* 12. doi, 10.1155/2008/790309.

Jimoh, K. O. & Kolapo, A. L. (2008). Mycoflora and aflatoxin production in market samples of some selected Nigerian foodstuffs. *Res. J. Microbiol. 3*, 169-174.

Jun, D. Y., Kim, J. S., Park, H. S., Song, W. S., Bae, Y. S. & Kim, Y. H. (2007). Cytotoxicity of diacetoxyscrpenol is associated with apoptosis by activation of caspase-(and interrution of cell cycle progression by down-regulating of cdk4 and cyclin B1 in human jurkat T cells. *Toxicology and Applied Pharmacology*, *222*, 190-201.

Jørgensen, K., Bak, S., Busk, P. K., Sørensen, C., Olsen, C. E., Puonti-Kaerlas, J. & Møller, B. L. (2005). Cassava Plants with a Depleted Cyanogenic Glucoside Content in Leaves and Tubers. Distribution of Cyanogenic Glucosides, Their Site of Synthesis and Transport, and Blockage of the Biosynthesis by RNA Interference Technology. *Plant Physiol. 139*, 363–374.

Kankolongo, M., Hell, K. & Nawa, I. N. (2009). Assessment for fungal, mycotoxin and insects spoilage in maize stored for human consumption in Zambia. *J. Sci. Food Agric. 89*, 1366-1375.

Kolind-Hansen, L. & Brimer, L. (2010). The retail market for fresh cassava root tubers in the European Union (EU); the case of Copenhagen, Denmark. A chemical food safety issue? *J. Sci. Food Agric. 90*, 252-256.

Kpodo, K., Sorensen, A. K. & Jakobsen, M. (1996). The occurrence of mycotoxins in fermented maize products. *Food Chem. 56*, 147-153.

Kurtzman, C. P. & Ciegler, A. (1970). Mycotoxin from a blue-eye mold of corn. *Appl. Microbiol. 20*, 204-207.

Lawley, R., Curtis, L. & Davis, J. (2008). Food safety harzard guidebook, *Royal Society of Chemistry*, 218.

Lewis, L., Onsongo, M., Njapau, H., Schurz-Rogers, H., Luber, G., Kieszak, S., Nyamongo, J., Backer, L., Dahiye, A. M., Misore, A., Decoct, K. & Rubin, C. (2005). Aflatoxin contamination of commercial maize products during an outbreak of acute aflatoxicosis in Eastern and Central Kenya. *Environ. Health Pers. 113*, 1763-1767.

Lomax, L. G., Cole, R. J. & Dorner, J. W. (1984). The toxicity of cyclopiazonic acid in weaned pigs. *Vet. Pathol. 21*, 418-424.

Lopez-Diaz, T. M., Santos, J. A., Garcia-Lopez, M. L. & Otero, A. (2001). Surface mycoflora of a Spanish fermented meat sausage and toxigenicity of Penicillium isolates. *Int. J. Food Microbiol. 68*, 69-74.

Lye, M. S., Ghazali, A. A., Mohan, J., Alwin, N. & Mair, R. C. (1995). An outbreak of acute hepatic encephalopathy due to severe aflatoxicosis in Malaysia. *Am. J. Trop. Med. Hyg. 53*, 68-72.

Matumba, L., Monjerezi, M., Chirwa, E., Lakudzala, D. & Mumba, P. (2009). Natural occurrence of AFB1 in maize and effect of traditional maize flour production on AFB1 reduction in Malawi. *Afr. J. Food Sci. 3*(12), 413-425

Manjula, K., Hell, K., Fandohan, P., Abass, A. & Bandyopadhyay, R. (2009). Aflatoxin and fumonisin contamination of cassava products and maize grain from markets in Tanzania and republic of the Congo. *Toxin Rev. 28*, 63-69.

McKinley, E. R., Carlton, W. W. & Boon, G. D. (1982). Patulin mycotoxicosis in the rat, Toxicology, pathology and clinical pathology. *Food Chem. Toxicol. 20*, 289-300.

Miles, D., Jansson, E., Mai, M. C., Azer, M., Day, P., Shadbolt, C., Stitt, V., Keirmeier, A. & Szabo, E. (2011). A survey of total hydrocyanic acid contant in ready-to-eat cassava-based chips obtained in the Australian market in 2008.

Mills, J. T. & Abramson, D. (1986). Production of sterigmatocystin by isolates of *Aspergillus versicolor* from Western Canadian stored barley and rapeseed canola. *Can. J. Plant Pathol. 8*, 151-153.

Mishkinsky, J. S., Goldschmied, A., Joseph, B., Ahronson, Z. & Sulman, F. G. (1974). Hypoglycaemic effect of Trigonella foenum graecum and Lupinus termis (leguminosae) seeds and their major alkaloids in alloxan-diabetic and normal rats. *Arch. Int. Pharmacodyn Ther. 210*(1), 27-37.

Mkumbira, J., Chiwona-Karltun, L., Lagercrantz, U., Mahungu, N. M., Saka, J., Mhone, A., Bokanga, M., Brimer, L., Gullberg, U. & Rosling, H. (2003). Classification of cassava into bitter and cool in Malawi, from farmers' perception to characterization by molecular markers. *Euphytica, 132*, 7-22.

Mokemiabeka, S., Dhellot, J., Kobawila, S. C., Diakabana, P., Ntietie Loukombo, R. N., Nyanga-Koumou, A. G. & Louembe, D. (2011). Softening and mineral content of cassava (Manihot esculenta Crantz) leaves during the fermentation to produce Ntoba mbodi. *Adv. J. Food Sci. Technol. 3*, 418-423.

Moorthy, S. N. & Mathew, G. (1998). Cassava fermentation and associated changes in physicochemical and functional properties. *Cri. Rev. Food Sci. Nutr. 38(2)*, 73-121.

Morant, A. V., Jørgensen, K., Jørgensen, C., Paquette, S. M., Sánchez-Pérez, R., Møller, B. L. & Bak, S. (2008). β-glucosidases as detonators of plant chemical defense. Phytochemistry 69:1795-1813. Mphande FA, Siame AB, Taylor JE. 2004. Fungi, aflatoxins and cyclopiazonic acid associated with peanut retailing in Botswana. *J. Food Prot. 67*, 96-102.

Muzanila, Y. C., Brennan, J. G. & King, R. D. (2000). Residual cyanogens, chemical composition and aflatoxins in cassava flour from Tanzanian villages. *Food Chem. 70*, 45-49.

Mwihia, J. T., Straetmans, M., Ibrahim, A., Njau, J., Muhenje, O. & Guracha, A. (2008). Aflatoxin levels in locally grown maize from Makueni District, Kenya. E. *Afr. Med. J. 85*, 311-317.

Nambisan, B. & Sundaresan, S. (1985). Effect of processing on the cyanoglucoside content of cassava. *J. Sci. Food Agric. 36*, 1197-1203.

Nartey, F. (1968). Studies on cassava, *Manihot utilissima* Pohl. I. Cyanogenesis, The biosynthesis of linamarin and lotaustralin in etiolated seedlings. *Phytochemistry* (oxf) *7*, 1307-1312.

Ngudi, D. D., Kuo, Y. H. & Lambein, F. (2003). Cassava cyanogens and free amino acids in raw and cooked leaves. *Food Chem. Toxicol. 41*(8), 1193-1197.

Nhassico, D., Muquingue, H., Cliff, J., Cumbana, A. & Bradbury, J. H. (2008). Rising African cassava production, diseases due to high cyanide intake and control measures. *J. Sci. Food Agric. 88*, 2043- 2049.

Nishie, K., Cole, R. J. & Dorner, J. W. (1987). Toxic effects of cyclopiazonic acid in the early phase of pregnancy in mice. *Res. Commun. Chem. Pathol. Pharmacol. 55*, 303-315.

Njobeh, P. B., Dutton, M. F., Koch, S. H., Chuturgoon, A., Stoev, S. & Seifert, K. (2009). Contamination with storage fungi of human food from Cameroon. *Int. J. Food Microbiol. 135*, 193-198.

Nuehring, L. P., Rowland, G. N., Harrison, L. R., Cole, R. J. & Dorner, J. W. (1985). Cyclopiazonic acid mycotoxicosis in the dog. *Am. J. Vet. Res. 46*, 1670-1676.

Nyirenda, D. B., Chiwona-Karltun, L., Chitundu, M., Haggblade, S. & Brimer, L. (2011). Chemical safety of cassava products in regions adopting cassava production and processing – Experience from Southern Africa. *Food and Chemical Toxicology, 49*, 607-612.

OECD. (2009). Consensus document on compositional considerations for new varieties of cassava (Manihos esculenta Crantz), Key food and feed nutrients, antinutrients, toxicants and allergens. Series on the Safety of Novel Foods and Feeds No. 18. 50 (ENV/JM/MONO(2009)44).

Obidoa, O. & Obasi, S. C. (1991). Coumarin compounds in cassava diets, 2 health implications of scopoletin in gari. *Plant Food Hum. Nutr. 41*, 281-289.

Oluwole, O. S. A., Onabolu, A. O., Mtunda, K. & Mlingi, N. (2007). Characterization of cassava (*Manihot esculenta* Crantz) varieties in Nigeria and Tanzania, and farmers' perception of toxicity of cassava. *J. Food Comp. Analysis*, *20*, 559-567.

Onabolu, A. O., Oluwole, O. S. A., Rosling, H. & Bokanga, M. (2002 a). Processing factors affecting the level of residual cyanohydrins in gari. *J. Sci. Food Agric. 82*, 966-969.

Onabolu, A. O., Oluwole, O. S. A. and Bokanga, M. (2002 b). Loss of residual cyanogens in a cassava food during short-term storage. *Int. J. Food Sci. Nutr. 53*, 343-349. Osuntokun, B. O. (1994). Chronic cyanide intoxication of dietary origin and a degenerative neuropathy in nigerians. *Acta Hortic.* (Wageningen) *375*, 311-321.

Oswald, I. P. & Comera, C. (1998). Immunotoxicity of mycotoxins. *Rev. Med. Vet. 149*, 585-590.

Padmaja, G. (1995). Cyanide detoxification in cassava for food and feed uses. *Cri. Rev. Food Sci. Nutr. 35*, 299-339. Pero, R. W., Harven, D., Owens, R. G. & Snow, J. P. (1972). A gas chromatographic method for the mycotoxin penicillic acid. *J. Chromatogr. 65*, 501-506.

Pfohl-Leszkowicz, A. (1999). La patuline. In, Tec and Doc (Eds), les mycotoxines dans l'alimentation, evaluation et gestion du risque (279-294). Paris, France. Conseil Superieur d'Hygiene Publique de France.

Prawat, H., Mahidol, C., Ruchirawat, S., Prawat, U., Tuntiwachwutticul, P., Tooptakong, U., Taylor, W. C., Pakawatchai, C., Skelton, B. W. & White, A. H. (1995). Cyanogenic and non-cyanogenic glycosides from *Manihot esculenta*. *Phytochemistry* (oxf) *40*, 1167-1173.

Purchase, I. F. & Van der Watt, J. J. (1969). Acute toxicity of sterigmatocystin to rats. *Food Cosmetics Toxicol. 7*(2), 135-139.

Purchase, I. F. (1971). The acute toxicity of the mycotoxin cyclopiazonic acid to rats. Toxicol. *Appl. Pharmacol. 18*, 114-123.

Rao, L. B. & Husain, A. (1985). Presence of cyclopiazonic acid in kodo millet (*Paspalum scrobiculatum*) causing "kodua poisoning" in man and its production by associated fungi. *Mycopathologia*, *89*, 177-180.

Ravindra, V. (1993). Cassava leaves as animal feed, potential and limitations. *J. Sci. Food Agric. 61*, 141-150.

Reijula, K. & Tuomi, T. (2003). Mycotoxins of Aspergilli. Exposure and health effects. *Front. Biosci. 8*, 232-235.

Salifu, A. (1981). Mycotoxins in short season sorghum in northern Nigeria. *Samaru J. Agric. Res. 1*, 83-88.

Saquib, M. (2000). A study of the impact of environmental requirements in India's export performance and the scoping study for environmentally friendly products, FAO, Mycotoxin prevention and control in food grains-aflatoxin research on grain in Asia, UNCTAD. http://www.fao.org/docrep/x0036e/x0036e1e.htm

Scudamore, K. A., Hetmanski, M. T., Clarke, P. A. & Barnes, K. A. (1996). Analytical methods for determination of sterigmatocystin in cheese, bread and corn products using HPLC with atmospheric pressure ionization mass spectrometric detection. *Food Addit. Contam. 13*, 343-358.

Seenappa, M., Keswani, C. L. & Kundya, T. M. (1983). *Aspergillus* infection and aflatoxin production in some cowpea (*Vigna unguiculata* (L.) Walp) lines in Tanzania. *Mycopathologia*, *83*, 103-106.

Siegers, C. P., Herzberg-Lottin, E., Von Otte, M. & Schneider, B. (1993). Anthranoid laxative abuse – a risk for colorectal cancer? *Gut*, *34*, 1099-1101.

Shephard, G. S. (2004). Mycotoxins worldwide, Current issues in Africa. In, Barug D, Van Egmond H, Lopez-Garcia R, Van Ossenbruggen T, Visconti A, (eds). Meeting the mycotoxin menace. Wageningen, *Wagningen Academic*. 81-88.

Shephard, G. S. (2008). Impact of mycotoxins on human health in developing countries. *Food Addit. Contam. 25*(2), 146-151.

Sosa, M. J., Cordoba, J. J., Diaz, C., Rodriguez, M., Bermudez, E., Asensio, M. A. & Nunez, F. (2002). Production of cyclopiazonic acid by *Penicillium commune* isolated from dry-cured ham on a meat extract-based substrate. *J Food Prot. 65*, 988-992.

Speijers, G. (1993). Cyanogenic Glycosides. *Food Add. Ser.* No. 30 Geneva: JECFA.

Speijers, G. J. A., Franken, M. A. M., Van Leeuwen, F. X. R., Van Egmond, H. P., Boot, R. & Loeber, J. G. (1986). *Subchronic oral toxicity study of patulin in the rat*. Report no. 618314 001. Rijksinstituut voor Volksgezondheid en Milieuhygiëne.

Steyn, P. S. & Rabie, C. J. (1976). Characterization of magnesium and calcium tenuazonate from *Phoma sorghina*. *Phytochemistry*, *15*, 1977-1979.

Sydenham, E. W., Vismer, H. F., Marasas, W. F. O., Brown, N. L., Schlechter, M. & Rheeder, J. P. (1997). The influence of deck storage and initial processing on patulin level in apple juice. *Food Add. Contam. 14*(5), 429-434.

Takahashi, H., Yasaki, H. & Nanayama, U. (1984). Distribution of sterigmatocystin and fungial mycelium in individual brown rice kernel naturally infected by *Aspergillus versicolor*. *Cereal Chem. 61*, 48-52.

Terao, K. (1983). Sterigmatocystin-masked potent carcinogenic mycotoxin. *J. Toxicol. Toxin Rev. 2*, 77-110.

Thorpe, C. W. & Johnson, R. L. (1974). Analysis of penicillic acid by gas-liquid chromatography. *J. Assoc. Off. Analyt. Chem. 57*, 861-865.

Tivana, L. D., Bvochora, J. M., Mutukumira, A. N. & Owens, J. D. (2007). A Study of heap fermentation process of cassava roots in Nampula Province, Mozambique. *J. Root Crops. 33*, 119-128.

Tylleskar, T. (1994). The association between cassava and the paralytic disease konzo. *Acta Hortic*. (Wageningen) *375*, 331-339.

Tylleskär, T., Howlett, W. P., Rwiza, H. T., Aquilonius, S.-M., Stalberg, E., Linden, B., Mandahl, A., Larsen, H. C., Brubaker,G. R. & Rosling, H. (1993). Konzo, a distinct disease entity with selective upper motor neuron damage. *Journal of Neurology, Neurosurgery, and Psychiatry*, *56*, 638-643

Valente-Soares L. M. & Rodriguez-Amaya D. B. (1989). Survey of aflatoxins, ochratoxin A, zearalenone and sterigmatocystin in some Brazilian foods by using multi-toxin thin-layer chromatographic method. *J. Assoc. Off. Anal. Chem. 72*, 22-26.

Van Egmond, H. P. (1983). Mycotoxins in dairy products. *Food Chem. 11*, 289-307.

Van Egmond, H. P. (2002). Worlwide regulations for mycooxins. *Adv. Exp. Med. Biol. 504*, 257-269.

Veringa, H. A., Van den Berg, G. & Daamen C. B. G. (1989). Factors affecting the growth of *Aspergillus versicolor* and the production of sterigmatocystin on cheese. *Neth. Milk Dairy J. 43*, 311-326.

Versilovskis, A., Bartkevics, V. & Miielsone, V. (2008). Sterigmatocystin presence in typical Latvian grains. *Food Chem. 109*, 243-248.

Versilovskis, A., Van Peteghem, C. & De Saeger, S. (2009). Determination of sterigmatocystin in cheese by high performance liquid chromatography-tandem mass spectrometry. *Food Addit. Contam. 26*, 127-133.

Wagacha, J. M. & Muthomi, J. W. (2008). Mycotoxin problem in Africa, current status, implications to food safety and health and possible management strategies. *Int. J. Food Microbiol. 124*, 1-12.

Wareing, P. W., Westby, A., Gibbs, J. A., Allotey, L. T. & Halm, M. (2001). Consumer preferences and fungal and mycotoxin contamination of dried cassava products from Ghana. *Int. J. Food Sci. Technol. 36*, 1-10.

Westby, A. (2002). Cassava utilization, storage and small-scale processing. In, RJ Hillocks, JM Tresh, AC Bellotti. (eds). *Cassava, Biology, Production and Utilization*. CAB International Publishing, Wallingford. pp. 281-300.

Westby, A., Wareing, P. W. & Gibbs, J. A. (1994). *Potential of cassava product to support mycotoxin formation*. In, Proceedings of the 10th International Symposium of the International Society of Tropical Root Crops Brazil, November 1994.

Wilkinson, H. H., Ramaswamy, A., Sim, S. C. & Keller, N. P. (2004). Increased conididation associated with progression along the sterigmatocystin biosynthesis pathway. *Mycologia. 96*, 1190-1(198.

Williams, J. H., Grubb, J. A., Davis, J. W., Wang, J., Jolly, P. E., Ankrah, N., Ellis, W. O., Afriyie-Gyawu, E., Johnson, N. M., Robinson, A. G. & Phillips, T. D. (2010). HIV and hepatocellular and esophageal carcinomas related to consumption of mycotoxin-prone foods in sub-Saharan Africa. *Am. J. Clin. Nutr. 92*, 1-7.

WHO. Joint FAO/WHO Expert Committee on Food Additives (JECFA). (1998). Position paper on patulin, 30th session, The Hague, The Netherlands.

Wu, F. (2006*). Mycotoxin reduction in Bt corn, potential economic, health, and regulatory impact*. ISB News Report, September (2006.

Yeoh., H.-H., &, Truong., V.-D. (1996). Protein contents, amino acid compositions and nitrogen-to-protein conversion factors for cassava roots. *J. Sci. Food Agric. 70*, 51-54.

Youssef, M. S., El-Maghraby, O. M. O., Ibrahim, Y. M. (2008). Mycobiota and mycotoxins of Egyptian peanut (*Arachis hypogeae* L.) seeds. *Int. J. Bot. 4*, 349-360.

Zasshi Y. (1963). Coumarin derivatives for medicinal purposes. XVII. pharmacological studies on coumarin derivatives having biological activity. *J. Pharm. 83*, 1124-1128.

In: Processed Foods
Editors: Chloe M. Gagne and Daniel B. Jones
ISBN: 978-1-62808-688-1

Chapter 4

TRANSFORMATION OF VEGETABLE WASTE OF THE FOOD INDUSTRY INTO VALUE-ADDED PRODUCTS: THE CASE OF *CYNARA SCOLYMUS*

Eliana N. Fissore[1,2], Cinthia Santo Domingo[1,3], Guadalupe Uricarriet[1], Elizabeth Erhardt[1], Ana M. Rojas[1,2] and Lía N. Gerschenson[1,2]*

[1]Industry Department,
School of Natural and Exact Sciences (FCEN), Buenos Aires University (UBA),
Ciudad Universitaria. Intendente Güiraldes 2620,
Ciudad Autónoma de Buenos Aires, Argentina
[2]Member of the National Scientific
and Technical Research Council-Argentina (CONICET)
[3]Fellow of the National Scientific
and Technical Research Council-Argentina (CONICET)

ABSTRACT

Byproducts of vegetable food processing represent a major disposal problem for the industry. Transformation of vegetable waste of the food industry into value added products as fibres are, may contribute to diminish pollution and to recover valuable biomass and nutrients.

A great number of diseases frequently occurring in developed countries such as various gastrointestinal disorders (hernia, duodenal ulcer, gall stones, Crohn's disease, appendicitis, constipation, hemorrhoids, colon carcinoma), diabetes mellitus, obesity and cardiovascular diseases, have a lower incidence among people consuming high amounts of fibre. The benefits of the intake of fibres for the gut are amply recognized, fact that has determined its consideration as a nutrient. Fibre addition to foods is an alternative to compensate for the existent deficiency in the diet.

---

* Corresponding author: Lia Noemi Gerschenson, e-mail: lia@di.fcen.uba.ar, Phone: 54 – 11 – 4576-3366 / 3397, Fax number: 54 – 11 – 4576-3366.

Besides the nutritional aspect, fibres can be used for technological purposes because they can change food consistency, texture, rheological behavior and sensory characteristics.

As only the tender inner bracts and the receptacle of the immature flowers of globe artichoke (*Cynara scolymus* L.) are used for consumption, residues can be upgraded to add value to its production, for example through their use for dietary fibre isolation.

In this research, it was evaluated a procedure for the obtention of fractions enriched in soluble fibre from artichoke stems. These tissues were dehydrated (80°C–2.5 h), milled, and dispersed in citrate buffer. Afterwards, cellulase was added and suspensions were incubated (30°C–20 h). Procedures were repeated but without enzyme. Soluble fibre fractions were precipitated with ethanol, filtrated, washed and freeze-dried. Yields obtained were 5% in the absence of enzymes and 7% when cellulase was used.

Fractions obtained were completely characterized observing a 69-70% (w/w) content of carbohydrates and a 25-28% (w/w) content of proteins. Carbohydrates included uronic acid and inulin. The use of the enzyme not only produced a yield increase but also an increase in uronic acid content.

Both fractions helped to stabilize oil-in-water emulsions probably due to their high protein content. The flow behavior of aqueous solutions of both fractions was pseudoplastic. The use of cellulase for fraction obtention, determined a lower consistency index and the existence of a minimum shear stress.

Procedure proposed allowed to isolate fractions enriched in soluble fibre, adding value to left over of *Cynara scolymus* industrialization, helping in the development of green processes for the agrifood chain and providing natural ingredients for more healthy foods.

INTRODUCTION

Due to a set of demographic and social changes, selling clean and cut fruits and vegetables for preparing soups and salads has expanded generating, along with traditional and emerging forms of industrialization, the existence of left over. According to Robert van Otterdijk and Alexandre Meybeck (2011), there are abundant food losses from agricultural production to food consumption. These losses imply a waste of resources used in production such as land, water, energy and supplies. In this way, the food production generates unnecessary CO_2 emissions and economic losses. The challenge ahead is to minimize these losses with different strategies, one of which may be the use of these residues to obtain value added products with sustainable procedures that help to reduce the pollution. In this way, changes in the environment will be minimized.

The benefits of fibres for the digestive system are widely recognized, fact that driven to its consideration as a nutrient (FDA, 1993). The matrix of the cell wall is the major source of fibre in fruits and vegetables. It is known that they are rich in pectic polysaccharides which have an important role in the diet. Thus, residues of the industrialization of fruits and vegetables might be used for obtaining soluble and insoluble fibres (i.e.: pectins, cellulose, hemicellulose), which can be applied to the food, pharmaceutical and cosmetic industries, ensuring full utilization and adding value to the raw material (Zambrano and Gallardo, 2002). The technique of isolation or fractionation of the fibres can affect their composition and functional characteristics. The properties can also be modified by technological treatments after obtention, which would determine a greater versatility in its application (Graham, 2002). Examples of the possible uses of isolated fibres are as texture modifier additives, as nutrient, as support for the development of lactic acid bacteria that produce metabolites of interest (Naveena et al., 2005), for the formation of edible films and coatings useful to improve the quality and lengthen the shelf life of food (Zambrano and Gallardo, 2002; Cherian et al., 2008; Buonocuore et al., 2003). Currently, special attention is paid to the isolation and characterization of fibres from non-traditional sources, due to the interest of scientists to evaluate their properties in relation to health and for the increase of the value added to the raw materials (Gerschenson et al., 2008). Depending on the source of the fibre and subsequent treatment, fractions with different compositions and physicochemical and functional characteristics are obtained (Fissore et al. 2007, 2009a, 2010; de Escalada Pla et al., 2007, 2010). Fissore et al. (2007, 2009a, 2010), working with pumpkin and beetroot, observed that tissue subjected to acid extraction resulted in a product containing galacturonic acid but with little thickening capacity and no effect on the diffusion of glucose. However, the enzyme treatment with hemicellulase, generated pumpkin pectin enriched products with capacity for delaying glucose diffusion and for acting as texture modifiers. In the food industry, pectins are used as thickeners, gelling agents and stabilizers, while new applications are constantly being developed. For example, Siew et al. (2008) reported that beet pectin has emulsifying properties.

Dietary Fibre

The current definition for the term dietary fibre (DF) was established in the year 2009, by the Codex Alimentarius Commission (Codex Dietary Fibre Definition. ALINORM 09/32/REP), which set the following parameters to classify the components that are defined as dietary fibre: DF means the carbohydrate polymers with ten or more monomeric units, which are not hydrolyzed by the endogenous enzymes of the small intestine of humans and belong to the following categories:

- edible carbohydrate polymers naturally occurring in foods in the form they are consumed;
- carbohydrate polymers obtained from food raw material by physical, enzymatic or chemical means and which have been demonstrated to have a physiological effect of benefit to health by generally accepted scientific evidence provided to the competent authorities;
- synthetic carbohydrate polymers which have been shown to have a beneficial physiological effect on health by generally accepted scientific evidence provided to the competent authorities.

Footnotes in the definition of dietary fibre in 2009 were modified in 2010 (ALINORM 10/33/26, 10/33/REP) and say:

- Dietary fibre, being of vegetal origin can include fractions of lignin and / or other compounds associated with polysaccharides in the plant cell wall and if such compounds are quantified by gravimetric analysis method proposed by AOAC for analyzing dietary fibre. These substances are included in the definition of fibre insofar as they are actually associated with the polysaccharide or oligo-saccharidic fractions of fibre. However, they may not be defined as DF if extracted and reintroduced into a food containing non-digestible polysaccharides. When combined with polysaccharides, these associated substances may provide additional beneficial effects.

- The decision on whether to include carbohydrates from three to nine monomeric units should be in the national authorities.

Types of Fibre

Plant tissues are sources of dietary fibre (DF). The cell wall components (ie: cellulose, lignin, non-cellulosic carbohydrates) are a source of dietary fibre (Selvendran and MacDougall , 1995). Dietary fibre can be classified into insoluble (IDF) and soluble (SDF), depending on their solubility in water. Another classification of the DF arise from its classification into two main types: (a) soluble / viscous / fermentable and (b) insoluble / non-viscous / slowly fermentable, according to Roberfroid (1993).

In food systems, dietary fibre is present in two forms: as a constituent of various plant foods (intrinsic) or added thereto. The aim of addition can be their use as ingredient (> 5%) or as an additive (<5%) according to Guillon and Champ (2000). The soluble fibre is comprised of pectins, some hemicelluloses (i.e. xylan, mannan, glucomannan, xyloglucan, glucans and arabinogalactan), fructans (i.e.: inulin), and other oligosaccharides. Soluble fibre has excellent qualities as a nutrient and as a food additive. Insoluble fibre consists mainly of cell wall components such as vegetable cellulose, lignin and certain hemicelluloses. Pulido and Saura Calixto (2000), informed that polyphenols associated with dietary fibre show antioxidant activity. Note that there are differences between the fibre extracted from fruits and vegetables and the one obtained from cereals. The total amount of soluble dietary fibre in fruits and vegetables is generally higher than in cereals (Figuerola et al., 2005). In terms of health benefits, both types of DF complement each other and a percentage of 70 - 50% of insoluble fibre and 30 - 50% of soluble fibre is considered to be a well-balanced proportion. Dietary fibre from cereal origin is a typical ingredient in products with high DF but the presence of soluble DF is low in cereals; for example, wheat bran contains 6% of soluble DF and 94 % of insoluble fibre. In contrast, for fruits and vegetables, the relation between soluble and insoluble DF is more balanced; for example, for pear , 39% of soluble and 61% of insoluble DF has been informed (Grigelmo-Miguel et al., 1999).

Pectin

Pectin is an abundant ubiquitous and multifunctional component of the cell wall of all land plants (Willats et al., 2006). Pectic polysaccharides consist mostly of polymers rich in galacturonic acid, containing significant amounts of rhamnose, arabinose and galactose as well as other 13 different monosaccharides (Vincken et al., 2003a,b). The three major pectic polysaccharides currently defined are homogalacturonan (HG), rhamnogalacturonan I (RG-I) and rhamnogalacturonan II (RG-II) (Waldron et al., 2003; Vincken et al., 2003a). Both the degree of methyl-esterification (DM) and the degree of acetylation (DA) have a profound impact on functional properties (Willats et al., 2006).

Pectins have been reported to lower serum cholesterol levels, to bind heavy metals,

and to have immunostimulating and anti-ulcer activities (Pellerin and O'Neill, 1998) and recently has been evaluated as an emerging new bioactive polysaccharide with anti-cancer effect (Maxwell et al., 2012). The ability of pectins to form gels determine its use in the food industry as gelling agents, thickeners and stabilizers (Waldron et al., 2003).

Inulin

Inulin has been defined as a polydisperse carbohydrate material consisting mainly, if not exclusively, of $\beta(2\rightarrow1)$ fructosyl-fructose. A starting glucose moiety can be present, but is not necessary (Franck and De Leenheer, 2005). The link $\beta(2\rightarrow1)$ is very special because it prevents inulin from being digested as a typical carbohydrate and is responsible for its low caloric value (Lopez-Molina et al., 2005). Inulin is found naturally in plants such as chicory root and artichoke, asparagus and onions (Zimeri and Kokini, 2002).

After starch, inulin is the reserve polysaccharide most abundant in nature, found in cereals, vegetables and fruits (Giannouli et al., 2004). It is usually found in the roots, tubers and rhizomes of certain flowering plants (i.e.: chicory, dandelion). Inulin extracted from different sources varies in its degree of polymerization (DP) which affects the functional attributes. If DP is high, then it is less soluble and is capable of forming micro-crystals which, although not noticeable in the mouth, can interact with food components to provide a smooth and creamy texture acting as a replacement for fat in breads, fillings and frozen desserts (Lopez-Molina et al., 2005). In fact, when used in cakes, pastries and dairy products at concentrations greater than 15%, inulin shows this skill (Glibowski and Glibowski and Glibowska, 2009).

The best known nutritional effect of inulin is the stimulation of intestinal bifidobacteria acting as substrate for its development, and therefore it is classified as a prebiotic (Lopez-Molina et al., 2005). Inulin acts synergistically with most gelling agents (i.e.: gelatin, alginate, kappa and iota carrageenan and maltodextrin). It also improves the stability of foams and emulsions, such as ice desserts and sauces (Franck and Coussement, 1997).

Glibowski (2010) showed that inclusion of inulin in concentrations of 20%, 25% and 30% in aqueous solutions determined an increase in viscosity. Inulin application as a gelling agent must be made with adequate control of thermal and mechanical conditions used during dissolution to prevent chemical changes. Moreover, in an acid medium, inulin can be hydrolyzed while at pH above 5 is stable and the gels obtained at near neutral pH are firm (Glibowski and Wasko, 2008).

Dietary Fibre Action in the Body

As mentioned above, the dietary fibre can not be absorbed and, therefore, reaches the large intestine intact. It is partially or completely fermented by intestinal bacteria (colon), causing gases and other compounds. Dietary fibre shortens digestion and helps to prevent constipation because as it is not being absorbed, adds bulk, increasing stool weight. Finally, the dietary fibre not being absorbed by the small intestine, in general, provides a low level of calories and therefore helps to control body weight, providing satiety.

However, the fraction fermented in the large intestine results in short chain fatty acids, which are responsible for providing energy value. During fermentation of the fibre, the short chain fatty acids generated are an energy source for the intestinal bacteria allowing their proliferation and their protective action can inhibit the development of cancer cells (Guillon and Champ, 2000).

On the other hand, soluble fibre, such as β-glucans and pectins, help to reduce the total cholesterol and LDL ("bad") cholesterol. So we can say that an adequate intake of fibre can help to prevent heart disease (Bundy et al., 2008).

Soluble fibre also helps to slow the absorption of glucose. The mechanism by which the soluble fibre reduces glucose level involves increasing the viscosity of food systems, thus hindering the absorption of glucose. Thus, through the consumption of these fibres, people with diabetes may have the option to better control glucose level, reducing their need for insulin (Guillon and Champ, 2000).

Incorporation of Fibre in Food

Adding fibre contributes to the modification and improvement of texture, modifying the sensory characteristics and the life of food products due to the ability of this component to absorb water and form gels. Soukoulis et al. (2009) conducted a study comparing the application of fibre extracted from apple. Adding apple fibre led to a greater viscosity increase (3 to 15 times greater than the control system), which could be attributed to the synergistic effect of soluble and insoluble fibre present. Gelroth and Ranhotra (2001) studied the physiological and nutritional aspects of dietary fibre leading to advise their incorporation in a large number of food products such as baked goods, breakfast cereals, baby food, meat products, pastries and yoghurts. Borderías et al. (2005) studied the application of fibre products to fish formulation observing the improvement of some functional properties such as gelling and water retention. From the nutritional point of view there was an increase in antioxidant capacity.

Techniques for Dietary Fibre Extraction

Fuentes-Alventosa et al. (2009) studied the effect of fibre extraction method on chemical composition and functional characteristics. The study was conducted on freshly harvested asparagus and the effect of factors such as treatment intensity (temperature - time), type of solvent and drying system were evaluated. The proportion of insoluble dietary fibre to soluble dietary fibre (IDF / SDF) decreased with the severity of the treatment, from 7.2 to 4.2 thus increasing the physiological quality of the fibre. The water retention capacity and the retardation of glucose absorption were higher in the heat treated fibre extracted. The water retention capacity showed values similar to those described for cereal fibres while the oil retention capacity was much higher.

Fissore et al. (2007, 2009b, 2010), working with pumpkin and beetroot also concluded that the procedure used for extraction of soluble fibre, had an important effect on fraction properties such as thickening effect. De Escalada Pla et al (2007, 2010) studied the effect of extraction method on the properties of fractions enriched in total dietary fibre obtained from quince and pumpkin and observed that density, specific volume, water retention capacity and oil holding capacity were conditioned by the raw material used and by the method used for fraction extraction.

Concerning the different techniques for soluble fibre extraction from the cell wall of fruits and vegetables, different techniques can be used, for example, acid treatment (pH 1.5) at 85 ° C for 1 hour (Robert et al., 2007) or the use of DEAE- cellulose chromatographic columns (Rombouts and Thibault , 1986). The purpose of these processes is to disassemble the structure of the cell wall, especially the cellulose microfibrils and thus release the chains of soluble fibre (i.e.: pectin or inulin.)

Pectin can be extracted from the cell wall material with water or buffer solutions, solutions of chelating agents, dilute acids, dilute sodium hydroxide (Garna et al., 2007). Many of these techniques mentioned are not industrially viable due to their low yield. In acidic medium, the solubility decreases and, therefore, it is a must to increase the extraction temperature to increase the yield.

In alkaline medium, some changes like the decrease of the degree of methylation, the acetylation degree and the length of the main chain, may occur, affecting the functionality of the product. In general, the extraction of pectin on an industrial scale is carried out in acidic conditions with pH between 1 and 3, temperatures between 50 and 100 ° C and an extraction time between 1 and 12 h, depending on the functional properties required (Levigne, 2002). Renard and Thibault (1993) reported that pectins resulting from extraction under industrial conditions are structurally more degraded than the product obtained through an extraction under mild conditions.

As for inulin extraction mode, Lopez-Molina et al. (2005) studied the extraction from the bracts of artichoke (*Cynara scolymus* L.) using only an aqueous solution at a temperature of 24°C for 12 h. With this method, fractions extracted had a high degree of polymerization. On the other hand, the most usual industrial extraction is performed from chicory roots (*Cichorium intybus*) given its high content of inulin, as well as from Jerusalem artichoke (*Helianthus tuberosus*) tubers and from ornamental plant *Dahlia single* L. (De Leenheer, 1996, Frutos et al., 2008).

Cellulase and Soluble Dietary Fibre

Cell wall has a structure in which pectin, cellulose, hemicellulose and protein are present. Cellulose acts as the main support for the structure. Cellulase is an enzyme capable of hydrolyzing cellulose into smaller units. By acting on the cell wall, the cellulase enzyme, produces a structural change of the wall, thus allowing the release of other components such as pectin (Fissore et al., 2009b). Cellulase is a complex enzyme containing endo-and exo-β-glucanase and cellobiase. Different cellulase preparations have different proportions of these components (Eveleigh et al., 2009).

Panouillé et al. (2006) compared the extraction of soluble fibre using cellulase and acid medium observing a higher yield with cellulase. Spagnuolo et al. (1997) studied the synergistic effects of cellulosic and pectic enzymes in the degradation of beet pulp.

Fan et al. (2010) patented a method for the bio-assisted isolation of inulin which involved the treatment of Jerusalem artichoke (*Helianthus tuberosus*) with cellulase and pectinase prior to extraction with water at a temperature of 40-55 ° C.

Globe Artichoke (*Cynara Scolymus* L.)

Artichoke (*Cynara scolymus* L.) is a perennial plant, widely cultivated for its large fleshy head or capitula (immature flowers) which represent 30-40% of the fresh weight. The edible parts of the artichoke are the tender inner bracts and the receptacle commonly known as "heart" that constitute nearly 35-55% of the fresh weight of the head. In addition, there are some non-food industrial by-products such as leaves, external bracts and stems (Christaki et al., 2012).

In recent years there has been a renewed and growing interest in the artichoke, an old plant with new applications in functional foods (Lattanzio et al., 2009). One of the ways in which it is industrialized is as canned artichoke. This industry generates large amounts of waste (stems, harder bracts). Lopez-Molina et al. (2005) and Lattanzio et al. (2009) reported that 70%-80% by weight of artichoke corresponds to waste. By having a composition similar to the edible part, this waste represents a promising source of value-added components: residues can be upgraded to add value to its production, for example through their use for dietary fibre isolation.

In the world about 120 000 ha are dedicated to artichoke growth and 90% of them are located in the countries of the Mediterranean basin. It was brought to America by italian immigrants and reached a remarkable development in the U.S.A. and Argentina (Castagnino, 2009), country that was since 2006 and until 2009, the fifth largest producer of artichoke (FAO, 2011).

The artichoke is a vegetable spread mainly in European countries. Italy is the world's largest producer followed by Spain. The 40% of Spanish artichokes are processed industrially. The autumn-winter crop is consumed fresh and the spring harvest is absorbed by the industry. Between 50 and 80% of European production is sold fresh and the rest, is industrialized. In the European Union (EU) fresh consumption has decreased but the industrialized product consumption in the form of hearts and pieces is important (Castagnino, 2009).

It is a medium-energetic value product (~40 Cal/100gr), is rich in minerals (K, Na, Ca, P, Fe),vitamins (A, B and C) (Tesi , 1990). The pharmacologic properties of artichoke flower heads are well documented in several in vivo and in vitro studies for the treatment of hepato-biliary dysfunction, dyspeptic syndromes, gastric diseases, as well as for inhibition of cholesterol biosynthesis and low density lipoproteins (LDL) oxidation -agents responsible for arteriosclerosis and coronary hearth disease (Lattanzio et al., 2009; Cristaki et al., 2012). According to Dreher (2001), the artichoke has a moisture content of~ 83 grams per 100 g of edible portion with a total fibre content of 7.6 grams, which can be divided into 4.4 grams of insoluble fibre and 3.2 grams of soluble fibre.

OBJECTIVE

The objective of this research was the study of mild techniques for extracting fractions rich in soluble fibres from artichoke (*Cynara scolimus* L.) stem tissue. For this purpose, the use of sodium citrate buffer solutions or of the buffer and the enzyme cellulase were explored. The chemical characterization of the isolated fractions and the study of their thickening and emulsifying effect in oil in water systems, was also a target of the investigation.

MATERIALS AND METHODS

Artichoke (*Cynara scolymus*) harvested in Argentina was bought in a local market. Hard bracts, hearts and stems were separated. Stems were washed with distilled water, dried (85 ºC, 2.5 h) in a convection oven (air rate: 0.508 m/s), milled (Wemir E909, Wemir, Buenos Aires, Argentina) and sieved for obtaining powders enriched in cell wall material (CWM) with particle sizes in the range 420-710 μm.

Preparation of Alcohol-Insoluble Residue and Its Characterization

In order to concentrate and characterize the polysaccharides present in the CWM, the alcohol insoluble residue (AIR) was obtained.

According to Martin-Cabrejas et al. (1994), 100.00 g of CWM was suspended in 350 ml of 85% (v/v)-ethanol, homogenized with a mechanical device and then boiled for 30 min under stirring. The residue obtained, was then extracted for 15 min with other 350 ml of 85% (v/v)-ethanol boiling solution. The last procedure was repeated another two times with 250 ml of 85% (v/v)-ethanol boiling solution, each procedure. The insoluble residue was separated and washed two times with 100 ml of 85% (v/v)-ethanol each time. Between each ethanol treatment, the suspension was filtered (Whatman filter: GF/C, UK) and the supernatant was discarded. The material obtained (AIR) was left overnight under a lab hood to eliminate the remaining ethanol, and then freeze-dried (Stokes freeze-drier, Stokes Company, Philadelphia, MA, USA) after freezing with liquid nitrogen. The product was then milled and the resulting powder was distributed in plastic tubes which were packaged in Cryovac (polyvinyl chloride-polyvinylidene chloride copolymer) bags under vacuum and stored at -28 °C till usage.

The AIR was used to evaluate total (non-cellulosic) carbohydrate, uronic acid, cellulose and lignin contents. Hydrolysis of cellulose and non-cellulosic polysaccharides of AIR was performed according to Ng et al. (1998) by dispersion of ~0.3000 g of sample product into 2080 μl of 72%-sulphuric acid.Solution (v/v) for 3 h at room temperature. This dispersion was made 1 M-sulphuric acid by addition of enough deionized water (until 25.00 ml-final volume) and each sample was heated at 100 °C for 2.5 h in a water-bath. After this, all dispersions were cooled, centrifuged at 12,000g for 10 min and the supernatant was separated, carefully neutralized and total carbohydrate content was determined by phenol–sulphuric method (Dubois et al., 1956). The residue was washed three times with deionized water, centrifuged at 12,000g for 10 min and, finally, freeze-dried. The residue obtained was weighed and reported as lignin.

A second procedure was carried out with other portion of ~0.3000 g of each sample dispersed into 2080 µl of 72%-sulfuric acid solution and water was immediately added up to 1 M-concentration followed by 2.5 h of heating at 100 °C. The final residue corresponded to cellulose + lignin.

The third hydrolysis-procedure was performed with a new portion of each sample following the technique applied for the second procedure, but 1 h of heating at 100 °C in a water-bath was applied in this case. Only the supernatant was separated for quantification as it was above indicated and uronic acid content was determined spectrophotometrically by the method reported by Filisetti-Cozzi and Carpita (1991).

Inulin and protein were also determined as described under "Methods for analysis of fractions".

All assays were performed in duplicate.

Extraction of Soluble Fibre from Artichoke

The powder enriched in cell wall material (CWM) artichoke obtained was digested in one liter of buffer solution (0.05M sodium citrate, pH 5.2) with 0.01% (w / w) sodium azide and 0.05 grams of cellulase enzyme per 10 grams of CWM. The hydrolysis was performed with constant stirring (10 rad s^{-1}) for 20 hours at 30 ° C. The system was filtered and two volumes of ethanol 96% (v / v) were added to the supernatant to isolate the soluble fibre. The system was filtered (Whatman filter: GF/C, UK) and the residue was lyophilized. With this procedure, the fraction C was obtained.

The enzyme cellulase C9422 (SIGMA, USA) from *Thricoderma viride* (activity: one unit liberates 1 mole of glucose from cellulose in one hour at pH 5.0 and 37 ° C) was used. The proportion of cellulase used in this work is consistent with previous studies by Fissore et al. (2009b) in tissues of beetroot and pumpkin, which showed that higher ratios of cellulase to substrate, increased the yield but were counterproductive to the functional properties of the products obtained.

Another extraction was conducted following the same steps as in the extraction previously described but without adding enzyme in order to evaluate the ability of sodium citrate buffer *per se* for soluble fibre obtention. The fraction obtained without enzyme is called fraction S.

Yield

Yield was calculated as g of product obtained per 100g of powder enriched in CWM used.

Methods for Analysis of Fractions

Chemical Analyses

All determinations were performed in triplicate using deionized water.

Deionized (Milli-Q^{TM}, USA) water was used for preparation of all reagents used. The following analyses were performed to each sample obtained:

a) Total carbohydrates by the phenol-sulphuric acid colorimetric method of Dubois et al. (1956).
b) Uronic acids through the spectrophotometric method reported by Filisetti-Cozzi and Carpita (1991).
c) Inulin content was determined using the fructan assay procedure for the measurement of fructo-oligosaccharides and fructan polysaccharides (AOAC Method 999.03; Megazyme, Ireland).
d) Protein content was determined into each sample by Lowry et al. (1951) assay using bovine serum albumin as standard (Sigma, St. Louis, USA).

Molecular Weight Profile

Samples were dialyzed against 0.5 M-imidazole buffer (pH 7.0) at 5 °C during 48 h, and then freeze-dried and dissolved in the imidazole buffer, as recommended by Mort et al. (1991). Molecular weight profiles were determined by gel filtration analysis using a Fast Protein Liquid Chromatograph (FPLC) with a Superose 12HR 10/30 column (Amersham, Pharmacia Biotech), eluted with 0.5 M-imidazole buffer (pH 7.0) at 0.5 ml/min (Wessels et al., 1998). Dextrans of 65000 Da and 40210Da molecular weights (SIGMA, St. Louis, USA), blue dextran (SIGMA, St. Louis, USA), $CoCl_2$ and sucrose were used as standards for column calibration.

Total carbohydrate content was determined into each collected fraction through the phenol-sulphuric acid method (Dubois et al.,1956).

Emulsifying Capacity

The concentration of fraction enriched in soluble dietary fibre used (2%, w/w) and the optimum time for emulsification (3 min) were selected in accordance with Williams et al. (2005).

Aqueous solutions were prepared by stirring each fraction in water to ensure complete dissolution and 10% of oily phase were added to constitute an oil in water emulsion. A middle-chain triglyceride (MCT) oil was used. The mixtures were subjected to shaking for 30 s immediately prior to homogenization for 3 min using an Ultra-Turrax mixer (IKA-WERKE, GMBH and Co. KG, Staufen, Germany) at 13,500 rpm. The emulsions were stored at room temperature and the droplet size was measured periodically and in triplicate, over 34 day storage using the Malvern Mastersizer 2000 (Malvern, UK). A few drops of the emulsion were added to water in the dispersion unit Hydro 2000MU of the instrument until the obscuration was ~15%.

The diameter of the drops was measured as the Sauter diameter (D_{32}) which is defined as:

$$D_{32} = \frac{\sum_i n_i d_i^3}{\sum_i n_i d_i^2}$$

where d_i is the droplet diameter and n_i is the number of droplets between two consecutive diameters (Jurado et al., 2007).

Rheological Characterization

The flow behavior of the samples was evaluated on aqueous systems containing

5% (w / v) of either fraction isolated in deionized water. Systems were submitted to stirring and allowed to stand for 24 h to assure complete hydration previous to characterization.

Studies were performed in a controlled stress rheometer (Paar Physica MCR 300, Anton Paar GmbH, Germany) equipped with 3 cm diameter serrated parallel plate geometry (PP30). A gap size of 500 μm was set and data points were recorded at steady state.

Each flow curve was determined in duplicate and in the same conditions: constant temperature of 25 ° C and strain rate range of 0.001 to 500 s^{-1}.

The following models were considered for fitting of data obtained.

Power Law

$$\tau = k \, . \, \dot{\gamma}^{n}$$

where τ represents the shear stress, k is the consistency index, is the strain rate and n is the flow index. If n = 1, the equation obtained is Newton's law and then, k = η (dynamic viscosity).

Herschel-Bulkley Model

$$\tau = \tau_0 + k \, \dot{\gamma}^{n}$$

where τ represents the shear stress, τ_0 represents the minimum stress necessary for the system to begin to flow, k is the consistency index, $\dot{\gamma}$ is the strain rate and n is the flow index.

The non-Newtonian fluids are characterized by their apparent viscosity η_a, which varies with strain rate. Thus, it can be said that:

$$\tau = \eta_a \cdot \dot{\gamma}$$

Therefore, the apparent viscosity can be calculated for a particular value of shear rate if k and n are known.

Color Measurement

The color of samples was assessed in triplicate by using a colorimeter (Minolta Co., Osaka, Japan) with illuminant D65 and 2° observer angle. Each sample was placed onto a white tile and values of CIE color space co-ordinates L, a and b values were acquired. L represents the luminosity (white to black), the parameter a represents the green (negative values) to red (positive values) axis and b represents the blue (negative values) to yellow (positive values) axis.

Measurements were performed in triplicate.

Data Modeling and Statistical Analysis

Non-linear fittings were performed using Prism 5 Statistical Software for Windows (GraphPad Software, Inc. La Jolla, CA, USA), as well as with the Solver function of the Excel program for Windows XP (Microsoft, Redmond, WA, USA). Statistical analyses for the comparison of results was carried out by use of ANOVA (level of significance, a: 0.05) followed by pairwise multiple comparisons using Tukey's significant difference test (Sokal and Rohlf, 2000).

RESULTS AND DISCUSSION

Cell Wall Material (CWM)

The yield of powder enriched in the cell wall material (CWM) was 12.1 g of powder per 100 g of stem tissue. The powder CWM was mainly composed of the cell wall polymers as indicated by its high proportion of alcohol insoluble residue (AIR) that was (67.2 ± 0.5) g per 100 g of CWM. AIR of the CWM consisted mainly of carbohydrates (76g/100g) with a high lignin content (21g/100g) as can be observed in Table 1. The calibration curve used for carbohydrate analysis was the fructose one, given the high content of inulin in the samples.

Wouters (2010) studied the inulin content of bracts and hearts of artichoke and reported a content of 3-10 grams of inulin per 100 grams of artichoke. Meyer et al. (2011) reported an inulin content of 1-5% (w/w) in fresh onions, between 4 and 12% (w/w) in fresh garlic, of 0.2 % (w/w) in banana and between 15 and 20% (w/w) in fresh chicory roots.

As seen in Table 5, the stem AIR showed a content of 20 g/100g cellulose, 8g/100g of uronic acid (pectin) and a high proportion of inulin (12 g/100g). The level of lignin obtained, 21 g/100 g AIR, is much higher than that obtained by Fissore et al. (2009b) in beetroot or pumpkin AIR (6.5 g/100 g and 2.8 g/100 g, respectively) fact that can be attributed to the histological and functional characteristics of the stem of the artichoke.

Table 1. Chemical composition of the alcohol-insoluble residue (AIR) of artichoke stem

COMPOSITION	Alcohol Insoluble ResidUE
Total carbohidrates[1](g/100g AIR)	76 ± 4
Uronic acid (g/100gr AIR)	8.0 ± 0.3
Cellulose (g/100g AIR)	20 ± 1
Inulin (g/100g AIR)	11.76 ± 0.02
Lignin (g/100g AIR)	21 ± 1
Protein (g/100g)	12± 1

Results are expressed as mean and standard deviation (n = 3).

[1] Expressed as g of fructose per 100 g of AIR (dry basis, db).

Characterization of Isolated Fractions

Yield and Chemical Analysis

Two different extraction processes were performed on the CWM, one using cellulase enzyme together with citrate buffer (C) and the other only with citrate buffer (S).

When the enzyme was used, a yield of 7 g/100 g CWM (db) was observed. In the absence of the enzyme, the yield was 5 g/100 g of CWM (db). It is important to remark that the salts of sodium citrate buffer may break the bonds in the cell wall, thus helping to isolate polysaccharides present (Fry, 1986).

Table 2 shows the composition of the fractions isolated from stem CWM. The content of total carbohydrates is similar for both fractions, 68.68 g of carbohydrates per 100 g of fraction C and 70.11 g of carbohydrates per 100 g of fraction S. The calibration curve used for this analysis was the fructose one, given the high content of inulin in the samples.

Uronic acid content was 18.52 g/100 g in fraction C and 12.05 g /100 g for fraction S. Probably the breakdown of cellulose by cellulase enzyme enabled to extract a greater proportion of cell wall pectin. The amount of inulin was 24.37 g/ 100 g in fraction C and 25.01 g/ 100 g for the fraction S. It is observed that the amount of inulin in these fractions is greater than the amount of pectin (galacturonic acid polymer).

A high protein content was observed for both fractions isolated: 27.79 g for fraction C and 25.33 g /100 g for fraction S. In the study conducted by López et al. (1999), it was determined that the amount of protein in the artichoke stem was 13.3 g of protein per 100 g of stem tissue on a dry basis. It can be concluded that the presence of enzyme increased the yield of fraction isolated. The greatest difference in the composition of both samples was observed for uronic acid content, having C a higher content than the fraction S. It must be also remarked that both fractions showed a high content of inulin and protein.

Table 2. Chemical characterization of the fractions isolated

COMPOSITION	FRACTION S	FRACTION C
Total carbohidrates (g/100g fraction, db)	70.11 ± 0.47	68.68 ± 0.15
Uronic acid (g/100g fraction, db)	12.05 ± 0.40	18.52 ± 0.74
Inulin (g/100g fraction, db)	25.01 ± 0.18	24.37 ± 0.38
Protein (g/100g fraction, db)	25.33 ± 1.60	27.79 ± 1.07

Results are expressed as mean and standard deviation (n = 3).

[1] Expressed as g of fructose per 100 g of fraction (dry basis, db).

Molecular Weight Profile

The molecular weight of a carbohydrate is very important because it defines certain functional properties such as solubility, ability to thicken or gel. Figures 1 and 2 show the molecular weight distribution for the fractions studied.

Different profiles are observed for both fractions. Regarding the sample obtained without enzyme (S) the following peaks can be observed in Figure 3: 260 kDa, 79 kDa, 17 kDa, 9480Da and 3820Da.

For the enzyme isolated sample (C), different peaks are observed in Figure 4: 890 kDa, 360 kDa , 107 to 17 kDa, 7000Da, 5100Da and 2800Da.

Profiles observed account for the heterogeneity of the fractions isolated, which is expected in the case of an unpurified sample comprising different polysaccharides and proteins.

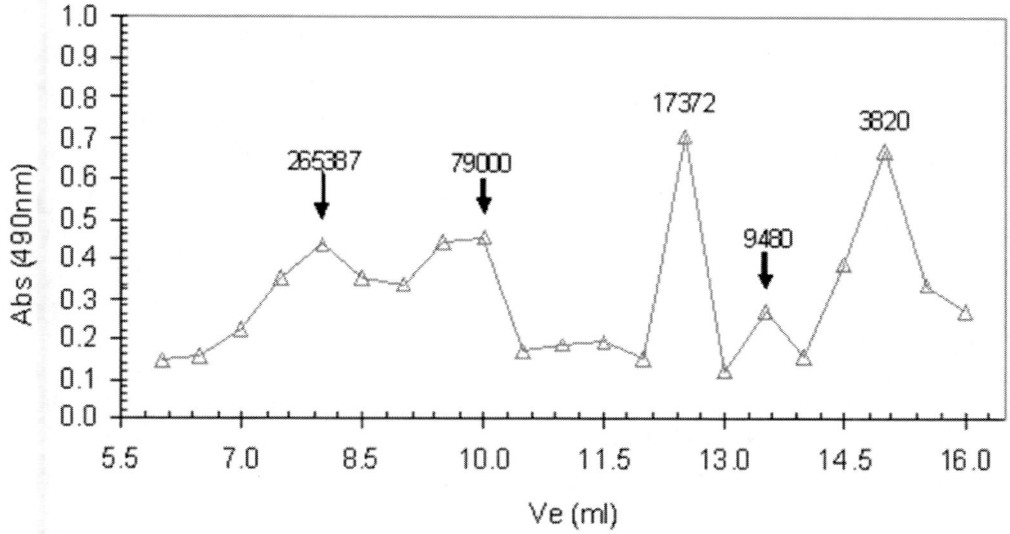

Figure 1. Distribution of molecular weights for fraction **S** isolated with buffer from stem artichoke.

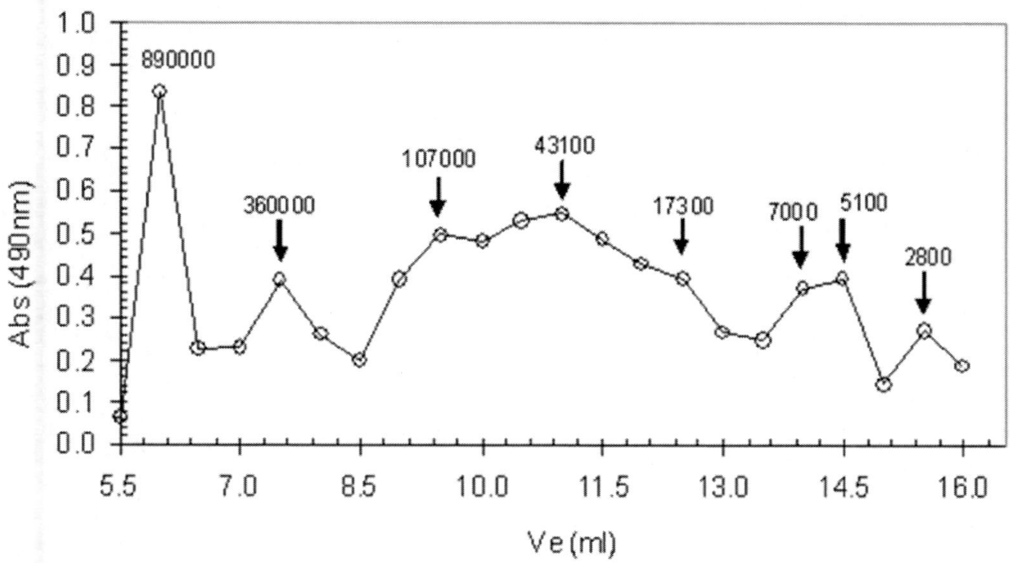

Figure 2. Distribution of molecular weights for fraction C isolated using cellulase and buffer from stem artichoke.

Westfall and Landis (1936) studied the molecular weight of inulin preparations using different temperatures and reported values ranging from 3300 Da to 5000 Da. Coulthard and Ruddock (1983) reported a value of 5000 Da. It is important to remark that according to the source of inulin, the degree of polymerization can vary widely from a few units to over 50 (Bot et al., 2003) and that the molecular weight values are affected by the method used for its determination. López-Molina et al. (2005) isolated inulin from artichoke and reported a degree of polymerization of 46 which corresponded to a highly polymerized inulin. Dan et al. (2009) used the intrinsic viscosity for molar mass determination and reported that the weight average molar mass of inulin monomer was 4468 Da and that they aggregate was found in aqueous solution with a weight average molar mass of 1.03×10^6 Da.

Williams et al. (2005) studied pectins obtained from sugar beet informing molecular weights ranging from 153 kDa to 562 kDa. Fissore et al. (2009 a and b) isolated pectin enriched fractions from pumpkin using cellulase/citrate buffer, informing molecular weights of 990 kDa and 740 kDa.

It can be concluded that fractions obtained showed high polydispersity, with high molecular weight polysaccharides that possibly corresponded to pectin fractions and peaks corresponding to low molecular weights (~5000 Da) that can be attributed to inulin.

Emulsifying Capacity

Funami et al. (2007) reported the emulsifying action of fractions rich in pectins isolated from sugar beet and attributed this effect to the protein present in the sample. Chanamai and McClements (2001) stated that carbohydrates can act as stabilizers of emulsions due to their thickening power. The stability of oil in water emulsions was studied based on the oil phase droplet size along 34 days. The effect of the presence of both fractions isolated at 2% (w / w) concentration was evaluated. The results are reported in Figure 3 where it can be observed that the emulsions maintained a relatively constant Sauter diameter during the 34-day trial. During the first week it was observed a slight increase of droplet size from about 1.5 to 1.9 μm and then, the emulsions remained stable.

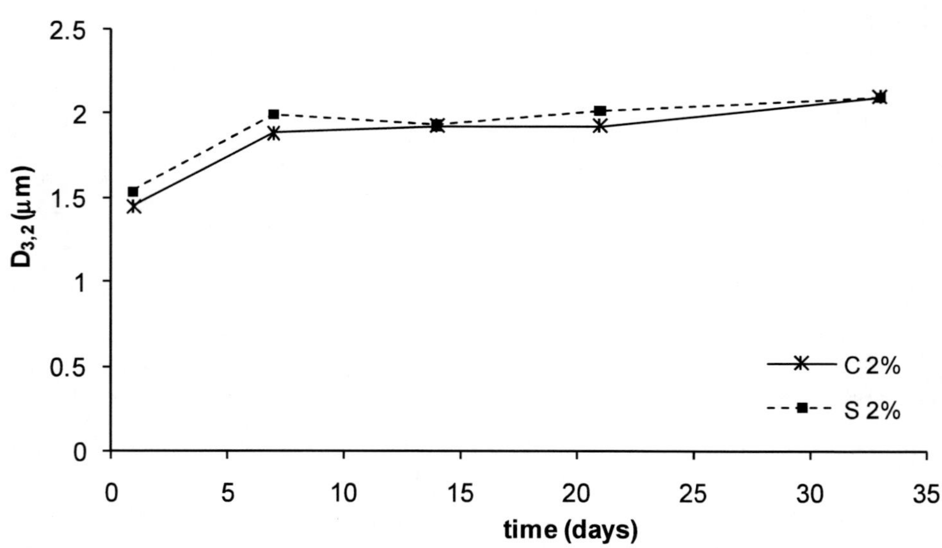

Figure 3. Stability of oil in water emulsions: change of oil droplet size with time. $D_{3,2}$: Sauter diameter.

The trend observed can be attributed to the increase of the continuous phase viscosity due to polysaccharide presence and to the reduction in the surface tension in the interface due to the relatively high level of protein present. Leroux et al. (2003) reported that the protein associated with the soluble fibre, which is protein of the primary cell wall rich in hydroxyproline, may play a key role in stabilizing the emulsions. It can be concluded that isolated fractions can help to stabilize the emulsions present in the food products to which they are incorporated and that these properties can also exert a beneficial effect in relation to the digestive processes.

Color

The color parameters L, a and b were evaluated.

Values observed in Table 3 indicate that the fractions obtained are dark and they have a yellow and red color. These characteristics might limit the use of isolated fractions in certain food products due to the effect on their visual appraisal but they can also be thought as natural and healthy conveyors of color when appropriate.

Frutos et al. (2008) studied the effect of artichoke fibre addition to wheat breads. They reported that lightness decreased with the increase of the amount of fibre incorporated.

Table 3. Color parameters

Muestra	L	a	b
S	26.3 ± 0.6	4.9 ± 0.1	9.3 ± 0.3
C	21.6 ± 1.1	4.1 ± 0.2	7.5 ± 0.4

Data are reported as mean and standard deviations.
S: fraction isolated without cellulase. C: fraction isolated with cellulase.
L: lightness; a: grade of greenness/ redness; b: grade of blue/yellow.

Rheological Characterization

Rheological characterization was performed in aqueous samples containing 5% (w / v) of each fraction.

Flow curves obtained are shown in Figures 4 and 5.

In Table 4, there are shown the parameters obtained from the fitting of experimental data to Power Law and Herschel-Bulkley model.

The experimental data for fraction S showed a suitable adjustment to power law resulting in a consistency index value (k) of 0.315 Pa s $^{0.711}$ and a flow index (n) of 0.711. Regarding the fraction obtained in the presence of cellulase (C), it showed an adequate fit to Herschel-Bulkley equation, with a consistency index (k) of 0.050 Pa s $^{0.881}$, a flow index (n) of 0.881 and a minimum shear stress (τ_0) of 0.260 Pa. The apparent viscosity values at 20 s^{-1} were 0.132 Pa s and 0.035 Pa s for the fraction S or C, respectively.

Fissore et al. (2009a) studied the flow behavior of aqueous systems containing a 2% (w/w) concentration of pectin-rich fractions isolated from pumpkin with citrate buffer, or cellulase /buffer and observed, under the same conditions used in the present work, a pseudoplastic behavior. Franck, A. and De Leenheer, L. (2002) reported apparent viscosity values (20 s^{-1}) of 0.0165 Pa s at 10 ° C for a 5% (w/w, dry matter) and of 1 Pa s for a 30% (w/w, dry matter) aqueous system based on inulin extracted from chicory stating that inulin can act as a thickener but at very high concentrations.

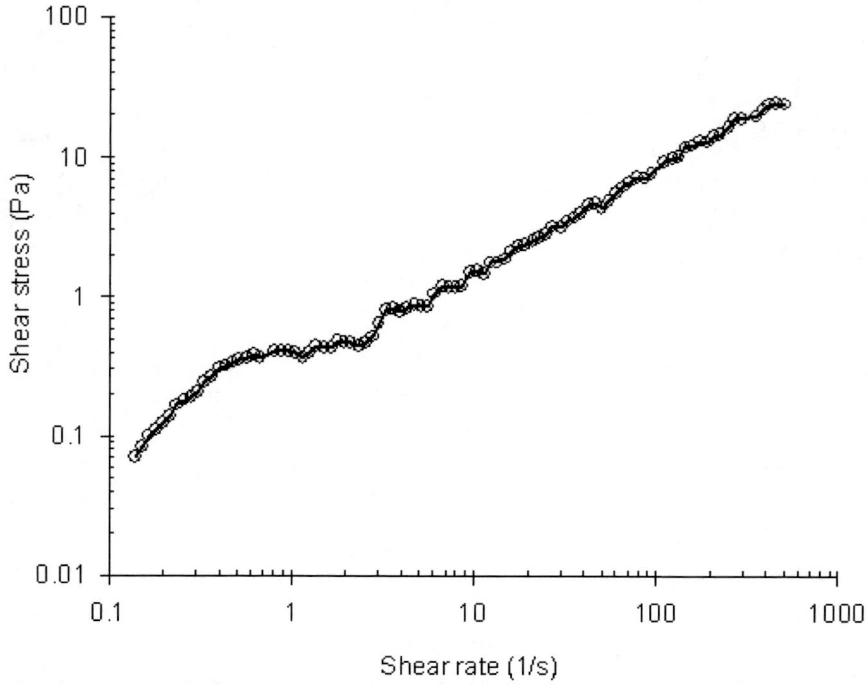

Figure 4. Flow curve at 25 ° C for aqueous system containing 5% (w / v) of fraction S isolated with buffer.

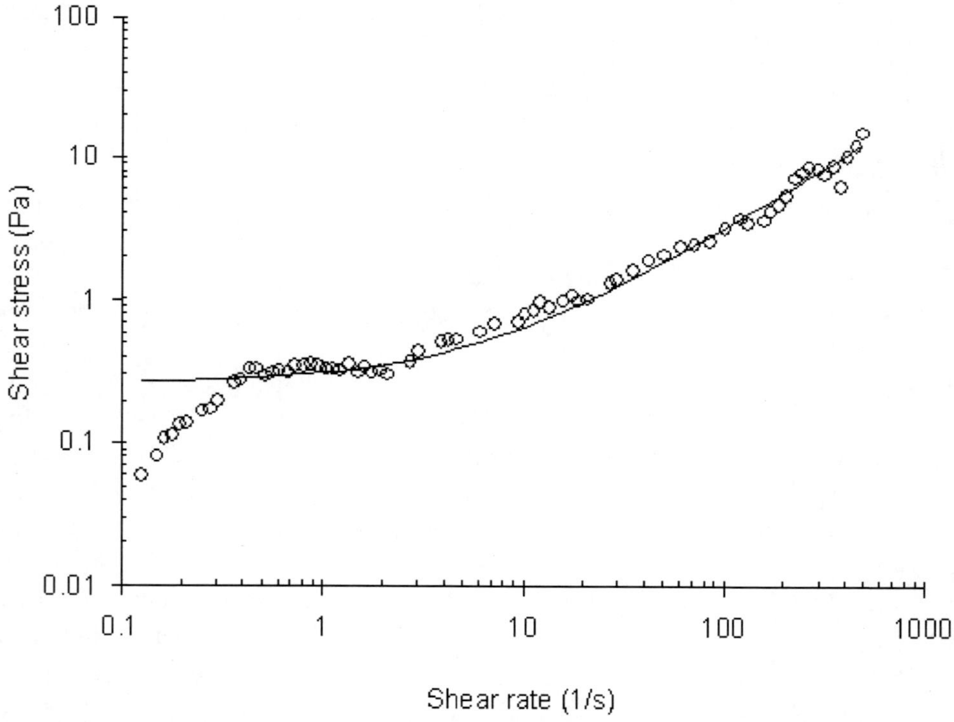

Figure 5. Flow curve at 25 ° C for aqueous system containing 5% (w / v) of fraction C isolated with cellulase.

Tabla 4. Apparent viscosity and flow parameters (25 °C) for aqueous systems containing the fractions isolated. Data are expressed as mean and standard error

Fraction	Apparent viscosity at 20 s^{-1} (Pa s)	τ_0 (Pa)	k (Pa s^n)	n	R^2	p
S	0.132 ± 0,006	ND	0.315 ± 0.013[1]	0.711 ± 0.008[1]	0.99	<0.0001
C	0.035 ± 0,009	0260± 0.075[2]	0.050 ± 0.011[2]	0.881 ± 0.037[2]	0.96	<0.0001

S: fraction isolated without cellulase. C. fraction isolated with cellulase. k: consistency index; n: flow index; τ_0 : minimum shear stress. R^2: coefficient of determination. p: probability. ND: not detected.
[1]Parameters were obtained from Power law.
[2]Parameters were obtained from Herschel-Bulkley model.

Hellwege et al. (2008) studied the thickening power of chicory inulin and found that a 20% (w/w) concentration in water gave origin to a viscosity of 2.2 mPa s, at 20 s^{-1} and 90 ° C. When inulin isolated from artichoke was used, in the same concentration in water, they observed a viscosity of 3 mPa s but when the concentration increased to 30% (w/w), the viscosity raised to 550 mPa s.

It can be concluded that although both fractions isolated showed pseudoplastic behavior, the soluble fibre enriched fraction isolated with citrate buffer from artichoke stem (fraction S), showed a slightly pronounced shear thinning behavior and a greater consistency index than the one isolated with citrate buffer and cellulase (C). The use of the enzyme for the extraction determined the existence of a minimum shear stress that can be useful from the technological point of view. The rheological behavior observed might be explained by the chemical composition of the samples which comprised pectin, inulin and proteins. In particular, the minimum shear stress observed for fraction C can be attributed to its higher concentration of uronic acid (Table 2).

CONCLUSION

Two procedures were assayed for the isolation of fractions enriched in soluble fibre (pectin and inulin) from stem of artichoke (*Cynara scolymus* L.). The yield obtained for the fraction isolated through a bio-assisted process using cellulase was slightly higher and this fraction showed a higher pectin content. The presence of inulin indicates the prebiotic capacity of these fractions. The fractions obtained showed low luminosity which might condition their application to food products but they can also be thought as natural and healthy conveyors of slight yellow/red color when appropriate.

Both fractions showed an emulsifying capacity for oil in water emulsions, probably due to their protein content. The flow behavior for both fractions was pseudoplastic and the fraction isolated in the presence of cellulase presented a minimum shear stress. These properties are important for technological applications and can also influence the digestive process.

In conclusion, it is shown that sodium citrate buffer and cellulase can be used for the obtention of fractions rich in soluble dietary fibre from the artichoke stem tissue, adding value to food industrialization waste. Products obtained can be used as healthy food ingredients.

REFERENCES

AOAC. (1991). Official Methods of Analysis, 15th ed. Washington, DC: Association of Official Analytical Chemists.

Borderías A.J., I. Sánchez-Alonso,I. and Pérez-Mateos,M. (2005). New applications of fibres in foods: Addition to fishery products. *Trends in Food Science and Technology* 16, 458–465.

Bot, A., Erle,U., Wreeker, R., Agterof, W. (2003) Influence of crystallisation conditions on the large deformation rheology of inulin gels. *Food Hydrocolloids* 18, 547 – 556.

Buonocore, G., Del Nobile, M., Panizza, A., Corbo, M. and Nicolais, L. 2003. *Journal of Controlled Release*, 90, 97-107.

Castagnino, A. M. (2009). Manual de Cultivos Hortícolas innovadores. Buenos Aires, Argentina: Editorial Hemisferio Sur.

Chanamai, R. and McClements, D. J. (2001). Depletion flocculation of beverage emulsions by gum arabic and modified starch. *Journal of Food Science*, 66, 457 – 463.

Cherian, B.M., Pothan, L.A., Nguyen-Chung, T., Mennig, G., Kottaisamy, M. and Thomas, S. (2008). *Journal of Agriculture and Food Chemistry*, 56, 5617-5627.

Christaki, E., Bonos, E. and Flourou-Paneri, P. (2012). Nutritional And Functional Properties of Cynara crops (Globe artichoke and Cardoon) and their potential applications: A review. *International Journal of Applied Science and Technology*, 2(2), 64-70).

Coulthard, M. and Ruddock, V. (1983). Validation of inulin as a marker for glomerular filtration in preterm babies. *Kidney International* 23, 407 – 409.

Dan A., Ghosh S. and Moulik S.P. (2009). Physicochemical studies on the biopolymer inulin: a critical evaluation of its self-aggregation, aggregate-morphology, interaction with water, and thermal stability. *Biopolymers,* 91(9), 687-699..

de Escalada Pla, M., Ponce, N., Stortz, C., Gerschenson, L.N. and Rojas. A.M.. (2007). Composition and Functional Properties of Enriched Fibre Products Obtained From Pumpkin (*Cucurbita moschata* Duchesne ex Poiret). *Lebensmittel Wissenschaft und Technology*, 40, 1176-1185.

de Escalada Pla, M., Uribe, M., Fissore, E., Gerschenson, L.N. and Rojas, A.M. (2010). Influence of the isolation procedure on the characteristics of fibre-rich products obtained from quince wastes. *Journal of Food Engineering*, 96(2), 239-248.

De Leenheer, L. (1996). Production and use of inulin: industrial reality with a promising future. In: H. Van Bekkum, H. Roper and F. Voragen (Eds.). Carbohydrates as Organic Raw Materials III (pp. 67–92). New York: VCH Publishing Inc.

Dreher, M. (2001). Dietary Fibre Overview. In: S. Sungsoo Cho (Eds.). Handbook of dietary fibre. (eBook ISBN: 978-0-203-90422-0; chapter 1). Boca Raton, Florida: CRC Press.

Dubois, M., Gilles, K.A., Hamilton, J.K., Rebers, P.A. and Smith, F. (1956). Colorimetric method for determination of sugars and related substances. *Analytical Chemistry*, 28, 350–356.

Eveleigh D., Mandels M., Andreotti R. and Roche C. (2009). Measurement of saccharifying cellulase. *Biotechnology for Biofuels*, 2-21. DOI:10.1186/1754-6834-2-21.

Fan, S., Liu, J., Wang, Z. (2010). Bio-Enzyme-Assisted Extraction Method for Inulin. US Patent Application No 2010/0119651 A1.

Figuerola, F., Hurtado, M. L., Estévez, A. M., Chiffelle, I. and Asenjo F. (2005). Fibre concentrates from apple pomace and citrus Peel as potencial fibre sources for food enrichment. *Food Chemistry*, 91(3), 395-401.

Filisetti-Cozzi, T. and Carpita, N.C. (1991). Measurement of uronic acids without interference from neutral sugars. *Analytical Biochemistry*, 197, 157 - 162.

Fissore, E., Matkovic, L., Wider, E., Rojas, A.M. and Gerschenson, L.N. (2009a). Rheological properties of pectins isolated from butternut (*Cucurbita moschata* Duch ex Poiret)". *Lebensmittel Wissenschaft und Technologie*, 42 (8), 1413-1421.

Fissore, E., Ponce, N., de Escalada Pla, M., Stortz, C., Rojas, A.M. and Gerschenson, L.N. (2010). Characterization of acid-extracted pectin-enriched products obtained from red beet (*Beta vulgaris* L. var. conditiva) and butternut (*Cucurbita moschata* Duch ex Poiret). 2010. *Journal of Agricultural and Food Chemistry,* 58(6), 3793-3800.

Fissore, E., Ponce, N., Stortz, C., Rojas, A.M. y Gerschenson, L.(2007). Characterisation of Fibre Obtained from Pumpkin (*Cucumis moschata* Duch.) Mesocarp Through Enzymatic Treatment. *Food Science and Technology International*, 13 (2), 141-151.

Fissore, E., Ponce, N., Wider, E.A., Stortz, C.A., Gerschenson, L.N. and Rojas, A.M. (2009b). Commercial cell wall hydrolytic enzymes for producing pectin-enriched products from butternut (*Cucurbita moschata* Duchesne ex Poiret). *Journal of Food Engineering*, 93, 293–301.

Food Agriculture Organization, FAO. (2011). FAOstat. http://faostat.fao.org/site/567/DesktopDefault.aspx?PageID=567#ancor.

Food and Drug Administration. FDA. Department of Health and Human Services. (1993). Health Claims: fruits and vegetables and cancer (Revised as of April 1, 2005). In: FDA, Code of Federal Regulations. Volume 2. Title 21. Chapter I. Subchapter B. Part 101. Subpart E. Section 101.78. Rockville, M.D.: US Government Printing Office.

Franck, A. and Coussement, P. (1997). Multifunctional inulin. *Food Ingredients and Analysis International,* 10, 8-10.

Franck, A. and Leenheer, L. (2002) Inulin. In: A. Steinbuchel (Ed.), Biopolymers, Vol. 6. (pp. 439-479). Weinheim, Germany: VCH.

Franck, A. and De Leenheer, L. (2005). Inulin. Biopolymers online (Wiley). DOI: 10.1002/3527600035.bpol6014.

Frutos, M.J., Guilabert-Antón, L., Tomás-Bellido, A. and Hernández-Herrero, J.A. (2008). Effect of artichoke (*Cynara scolymus* L.) fibre on textural and sensory qualities of wheat bread. *Food Science and Technology International* , 14 (5), 49-56.

Fry S.C. (1986). Cross-linking of matrix polymers in the growing cell walls of angiosperms. *Annual Review of Plant Physiology*, 37, 165 – 186.

Fuentes- Alventosa, J.M., Rodríguez-Gutiérrez, G., Jaramillo-Carmona, S., Espejo-Calvo, J.A., Rodríquez-Arcos, R., Fernández-Bolaños, J., Guillén-Bejarano, R. and Jiménez-Araujo, A. (2009). Effect of extraction method on chemical composition and functional characteristics of high dietary fibre powder obtained from asparagus by-products. *Food Chemistry* 113, 665 - 671.

Funami, T., Zhang, G., Hiroe, M., Noda, S., Nakauma, M. and Asai, I. (2007). Effects of the proteinaceous moiety on the emulsifying properties of sugar beet pectin. *Food Hydrocolloids*, 21, 1319 - 1329.

Garna, H.B., Mabon, N., Robert, C., Cornet, C., Nott, K., Legros, H., Wathelet, B. and Paquot, M. (2007). Effect of extraction conditions on the yield and purity of apple pomace pectin precipitated but not washed with alcohol. *Journal of Food Science*, 72(1), C1 - C9.

Gelroth, J. and Ranhotra, G.S. (2001). Food uses of fibre. In: S. Sungsoo Cho (Eds.). Handbook of dietary fibre. (eBook ISBN: 978-0-203-90422-0; chapter 23). Boca Raton, Florida: CRC Press.

Gerschenson, L. N., Rojas, A. M., de Escalada Pla, M. and Fissore, E. 2008. Functional properties of dietary fibre isolated from *Cucurbita moschata* Duchesne ex Poiret through different extraction procedures. In J.N. Govil and V.K. Singh (Eds.). Recent Progress in Medicinal Plants. Vol. 24. (pp. 359-370). Houston, Texas: Studium Press Llc.

Giannouli P., Richardson R. K. and Morris E. R. (2004). Effect of polymeric co-solutes on calcium pectinate gelation: dextrans and inulin. *Carbohydrate polymers*, 55, 357 - 365.

Glibowski, P. (2010). Effect of thermal and mechanical factors on rheological properties of high performance inulin gels and spreads. *Journal of Food Engineering*, 99, 106 – 113.

Glibowski P. and Glibowska A. (2009). Effect of calcium chloride on rheological properties and structure of inulin-whey protein gels. *International Journal of Biological and Life Sciences,* 5(3), 134-138.

Glibowski, P. and Wasko, A. (2008). Effect of thermochemical treatment on the structure of inulin and its gelling properties. *International Journal of Food Science and Technology*, 43, 2075 – 2082.

Graham, B.S. (2002). Pectins and their manipulation. Londres, UK:Blackwell Publishing Co.

Grigelmo-Miguel, N., Gorinstein, S. and Martín-Belloso, O. (1999). Characterization of peach dietary fibre concentrate as a food ingredient. *Food Chemistry*, 65, 175 -181.

Guillon, F. and Champ, M.M.J. (2000). Structural and physical properties of dietary fibres, and consequences of processing on human physiology. *Food Research International*, 33, 233 - 245.

Hellwege, E., Peeters, R. and Pilling, J. (2008). Long-chain inulin. US Patent Application No: 2008/0255249.

Jurado, E., Bravo, V., Camacho, F., Vicaria, J.M. and Fernandez-Arteaga, A. (2007). Estimation of the distribution of droplet size, interfacial area and volume in emulsions. *Colloids and Surfaces A: Physicochemical Engineering Aspects*, 295, 91-98.

Lattanzio, V., Kroon, P., Linsalata, V. and Cardinali A. (2009). Globe artichoke: a functional food and source of nutraceutical ingredients. *Journal of functional foods*, 1, 131-144.

Leroux, J., Langendorff, V., Schick, G., Vaishnav, V. and Mazoyer, J. (2003). Emulsion stabilizing properties of pectin. *Food Hydrocolloids*, 17, 455 - 462.

Levigne, S., Ralet, M.-C. and Thibault, J.-F. (2002) Characterization of pectins extracted from fresh sugar beet under different conditions using an experimental design. *Carbohydrate Polymers*, 49, 145 - 153.

Lopez, G., Ortuño, J., Periago, M.J., Martínez, C. and Rincón, F. (1999). Influencia del tratamiento térmico y la fibra dietética en la calidad de la proteína de la alcachofa y su subproducto. *Archivos latinoamericanos de Nutrición*, 49 (1), artículo 10.

Lopez-Molina, D., Navarro-Martínez, M.D., Rojas Melgarejo, F., Hiner, A., Chazarra, S., and Rodríguez-Lopez, J.N. (2005). Molecular properties and prebiotic effect of inulin obtained from artichoke (*Cynara scolymus* L).

Lowry, O.H., Rosebrough, N.J., Farr, A.L. and Randall R.J. (1951). Protein measurement with the Folin phenol reagent. *Journal of Biological Chemistry*, 193, 265-275.

Martin-Cabrejas, M.A.M., Waldron, K.W. and Selvendran, R.R. (1994). Cell wall changes in spanish pear during ripening. *Journal of Plant Physiology*, 144, 541-548.

Maxwell E. G., Belshaw N., Waldron K. W. and Morris V. J. (2012) Pectin - *An emerging new bioactive food polysaccharide Trends in Food Science and Technology*, 24 (2), 64-73.

Meyer, D., Bayarri, S., Tárrega, A. and Costell E. (2011). Inulin as texture modifier in diary products. *Food Hydrocolloids*, 25, 1881-1890.

Mort, A.J., Moerschbacher, B.M., Pierce, M.L. and Maness, N.O.(1991). Problems encountered during the extraction, purification, and chromatography of pectic fragments, and some solutions to them. *Carbohydrate Research*, 215, 219-227.

Naveena, B.J., Altaf, M., Bhadriah, K. and Reddy, G. (2005). *Bioresource Technolology*, 96, 485-490.

Ng, A., Parr, A.J., Ingham, M.L., Rigby, N.M. and Waldron, K.W. (1998). Cell wall chemistry of carrots (*Daucus carota* L.,cv. Amstrong) during maturation and storage. *Journal of Agricultural and Food Chemistry*, 46, 2933-2939.

Panouillé, M., Thibault, J.F., Bonnin, E. (2006). Cellulase and protease preparations can extract pectins from various plant byproducts. *Journal of Agricultural and Food chemistry*, 54, 8926 -8935.

Pellerin, P. and O'Neill, M.A. (1998). The interaction of the pectic polysaccharide rhamnogalacturonan II with heavy metals and lanthanides in wines and fruit juices. *Analitycal Magazine*, 26, 32-36.

Pulido, R., Bravo, L. and Saura-Calixto, F. (2000). Antioxidant activity of dietary polyphenols as determined by a modified ferric reducing/antioxidant power assay. Journal of Agricultural and Food Chemistry, 48(8), 3396-402.

Renard, C. and Thibault, J. (1993). Structure and properties of apple and sugar-beet pectins extracted by chelating agents. *Carbohydrate Research*, 244, 99 – 114.

Roberfroid, M. (1993). Dietary fibre, inulin, and oligofructose: a review comparing their physiological effects. *CRC Critical Reviews in Food Science and Nutrition*, 33, 103-148.

Robert, C., Happi Emaga, T., Wathelet, B. and Paquot M. (2007). Effect of variety and harvest date on pectin extracted from chicory roots (*Cichoiurm intybus* L.). *Food Chemistry*, 108, 1008 – 1018.

Rombouts, F. and Thibault, J. (1986). Sugar beet pectins: chemical structure and gelation through oxidative coupling. In: M.L. Fishman and J. Jen (Eds.), Chemistry and function of pectins (pp. 49-60). Washington, DC: American Chemical Society.

Selvendran, R.R. and MacDougall, A.J. (1995). Cell-wall chemistry and architecture in relation to sources of dietary fibre. *European Journal of Clinical Nutrition*, 49(3), 27-41.

Siew, C. and Williams, P.A. (2008). Role of protein and ferulic acid in the emulsification properties of sugar beet pectin. Journal of Agricultural and Food Chemistry*, 56(11), 4164-4171.*

Sokal, R. R. and Rohlf, J. B. (2000). Biometry. The principles and practice of statistics in biological research. San Francisco,CA: WH Freeman and Company.

Soukoulis, C., Lebesi, D. and Tzia, C. (2009). Enrichment of ice cream with dietary fibre: effects on rheological properties, ice crystallisation and glass transition phenomena. *Food Chemistry*, 115, 665- 671.

Spagnuolo, M., Crecchio, C., Pizzigallo, M. and Ruggiero, P. (1997). Synergistic effects of cellulolytic and pectinolytic enzymes in degrading sugar beet pulp. *Bioresource Technology,* 60, 215- 222.

van Otterdijk, R. and Meybeck, A. (2011). Global food losses and food waste. *Extent, causes and production.* Rome, Italy: FAO.

Vincken, J. P., Schols, H., Oomen, R. J. F. J., Beldman, G., Visser, R. and Voragen, A. G. J. (2003a). Pectin-the hairy thing. In: A. G. J. Voragen, H. Schols and R. Visser (Eds.), Advances in pectin and pectinase research (pp. 47–59). Dordrecht, Netherlands: Kluwer Academic Publishers.

Vincken, J. P., Schols, H. A., Oomen, R. J. F. J., McCann, M. C., Ulvskov, P., Voragen, A. G. J. and Visser, R. G. F. (2003b). If homogalacturonan were a side chain of rhamnogalacturonan I. Implications for cell wall architecture. *Plant Physiology*, 132, 1781–1789.

Waldron, K. W., Parker, M. L., and Smith, A. C. (2003). Plant cell walls and food quality. *Reviews in Food Science and Food Safety*, 2, 101-119.

Wessels, M.R., Paoletti, L.C., Guttormsen, H.K., Michon, F., D'Ambra and A.J., Kaspe, D.L. (1998). Structural properties of group B streptococcal type III polysaccharide conjugate vaccines that influence immunogenicity and efficacy. *Infection and Immunity*, 66(5), 2186-2192.

Westfall, B. and Landis, E. M. (1936). The molecular weight of inulin. *The Journal of Biological Chemistry*, 116, 727-734.

Williams, P.A., Sayers, C. ,Viebke, C. , Senan, C., Mazoyer, J. and Boulenguer, P. (2005). Elucidation of the Emulsification Properties of Sugar Beet Pectin. *Journal of Agricultural and Food Chemistry, 53* (9), 3592-3597.

Willats, W.T., Knox, J.P. and Mikkelsen, J.D. (2006). Pectin: new insights into an old polymer are starting to gel. *Trends in Food Science and Technology*, 17 (3), 97-104.

Wouters, R. (2010). Inulin. In: A. Imeson (Ed.), Food stabilisers, thickeners and gelling agents. (pp. 180-197). Chichester, UK: Wiley-Blackwell.

Zambrano-Zaragoza M. and Gallardo-Navarro Y. (2002). Temas en Tecnología de Alimentos, Fibra dietética. Mexico: Editorial Alfaomega.

In: Processed Foods
Editors: Chloe M. Gagne and Daniel B. Jones

ISBN: 978-1-62808-688-1

Chapter 5

FUNCTIONAL MATHEMATICAL INDEX (FMI): AN INDEX GENERATOR FOR "TAMING" QUALITY, APPLIED TO FOOD AND PROCESSES

Enrico Finotti[1,*], Alberto Bersani[2,3], Enrico Bersani[3] and Ernesto Del Prete[4]

[1]Istituto Nazionale di Ricerca per gli Alimenti e la Nutrizione, Rome, Italy
[2]Dipartimento SBAI, Sapienza Università di Roma, Rome, Italy
[3]Laboratorio di Strutture e Materiali Intelligenti - Sapienza Università di Roma sede distaccata di Cisterna di Latina, Cisterna di Latina LT, Italy
[4]INAIL, Rome, Italy

ABSTRACT

In several fields, e.g. in Nutritional Sciences and Food Technology, it is important to establish the right way to detect and define the quality of food and related processes.

Generally speaking the quality depends not only on the choice of measurable parameters, allowing to determine food quality in an objective way, but also on non "scientific" and nonmeasurable aspects.

In a general way useful parameters involved in quality determination are not comparable to each other (i.e. different measure units and scales). In addition, in literature several quality measurement techniques make use of weights - associated to the characteristic parameters - which introduce a sort of subjectivity in defining quality.

In order to solve these methodological problems, for our purposes we developed a Functional Mathematical Index (FMI) that allows to define the quality in food and processes without weights. FMI is a number obtained by a suitable choice of a set of parameters that in a unique way characterizes the quality of food and processes. The choice of number and type of parameters allows to make some particular features to raise up.

---

*Corresponding author: Enrico Finotti Ph.D., INRAN Via Ardeatina 546, 00178 – Rome, Italy, e-mail: finotti@inran.it, Tel.: +39 06 51494430, Fax: +39 06 51494550.

Mathematically speaking, FMI was introduced as the norm of a vector in an N-dimensional space, where N is the number of chosen parameters. The chosen parameters are normalized (their values are between 0 and 1), and summed together. Further studies of our research group have modified and improved the mathematical meaning of FMI, which however can be interpreted, in a very general way, as the distance from the "ideal" food and the considered sample. The better the quality is, the closer to 0 FMI is. The core of FMI building process is the appropriate choice of the parameters.

We tested FMI with good results on different products, e.g. extra virgin olive oil along the ripening phase, black and green tea, several potato cultivars cooked in many ways, durum wheat etc.

The FMI shows a wide ductility and adaptability to a wide range of foods and to different application fields.

In order to evaluate the importance and the usefulness of FMI, in this review we will summarize the development of this index and we will show the last definition applied onto food and technical processes, with numerical examples.

INTRODUCTION

In several fields, e.g. in Nutritional Sciences and Food Technology, it is important to establish the right way to detect and define the quality of food and related processes.

In general, the first step is considering food composition and quantifying principal nutrients, that can be distinguished into four different classes:

- Macronutrients, i.e., nutrients needed in considerable amount such as carbohydrates, proteins, fats.
- Micronutrients, i.e., nutrients needed in small quantity such as minerals and vitamins.
- Essential Nutrients, i.e., compounds that can not be synthesized at all or just partially by human body for its needs.
- Dietary fibres, i.e., all materials that can not hydrolysed by enzymatic reactions during the digestive processes, but are transformed by microflora.

MEASURABLE PARAMETERS

Carbohydrates

The most important carbohydrates for human diet are free glucose and fructose that are the only carbohydrates that can be absorbed by the intestinal tract. These carbohydrates come from single molecules or from polymeric carbohydrates (polysaccharides), and from enzymatic reactions metabolized into single units. The most important polysaccharide for human diet is starch (homopolysaccharide) that is made up of only one single type of molecule, D-glucose. This carbohydrate is very important in human diet because it keeps the glucose level constant in blood by enzymatic metabolism. In food technologies D-glucose represents a sugar with functional properties during the processes (gelatinization etc.) [1, 2].

Proteins

Proteins are polymers of amino acids and can be present in four different forms. All living tissues have proteins: human tissues need a total of twenty amino acids. Some of them are not essential amino acids: alanine, arginine, asparagine, aspartic acid, glutamic acid, glutamine, glycine, ornithine, proline, and serine, while the others are essential for human life: histidine, isoleucine, leucine, lysine, methionine+ cystine, phenylalanine+ tyrosine, threonine, tryptophan and valine. The presence of these last eight essential amino acids must be taken into account in every diet; the lack of one of these essential amino acids makes it "limiting amino acid" and every nutritional table of aliments reports for each food its own limiting amino acid.

The abundance (amino acid score) of these amino acids gives us the nutritional value of each single protein. This value is calculated by FAO amino acids score [3].

Lipids

Lipids are another important class of nutrients: triglycerides, sterols, waxes, phospholipids, etc.. The importance of triglycerides is principally due to their fatty acids composition. In fact the presence of essential fatty acids, such as linoleic acid (C18:2, n-6) and α-linolenic acid (C18:3, n-3), increases the nutritional value of food. From a nutritional point of view, also the presence and the position of double bounds in the fatty acid carbon atoms chain makes these compounds newsworthy. In particular the balance of double bounds in n-3 and in n-6 position along the fatty acid carbon atoms chain, or the presence of cis/trans fatty acids is important [4]. β-sitosterol is the principal sterol present in vegetable foods. It can play an interesting role in the cholesterol diet, because it reduces the feed cholesterol absorption.

Cholesterol and phospholipids are indispensable because they play a crucial role in the muscle cell membranes [5].

Vitamins

Vitamins can be divided into two classes: water soluble vitamins such as thiamine, riboflavina, vitamin B_6, vitamin B_{12}, folate, pantothenic acid, biotin and vitamin C, and fat soluble vitamins, such as vitamin A, vitamin D, vitamin E and vitamin K. These compounds perform a wide range of physiological activities, as co-factors in enzyme activity or as antioxidants or as pro-hormones. Their deficiency could be very dangerous for human health, but these compounds are double sided. Hypervitaminosis, an overdosage of vitamins, leads to intoxication [2, 6].

Minerals

Inorganic minerals such as calcium, magnesium, phosphorus, sodium, potassium, chloride, iron, zinc, copper, selenium, molybdenum, manganese, chromium, iodine, fluoride and silicon, are involved into different and important physiological functions and, also in this case, their dosage is very important, because each mineral has its own range of concentration (maximum and minimum value) where they have a physiological role. Out of their bounds they can produce different dysfunctions in human body [7].

In this short review, about the principal nutrients that human diet requires by food, we have deliberately stressed not only the common and well know properties and their energetic value, but also their peculiarities, because, at the end, these peculiarities have a great impact on the human health. In any diet the absolute presence of carbohydrates is not so important as the possibility to obtain, by enzymatic reactions, glucose and fructose; the presence of proteins is not so important as the presence of amino acids, in particular the essential amino acids; the presence of triglycerides is not so important as the kind of fatty acids they contain; the presence of vitamins and minerals in food is not so important as their dosage.

Antinutrients

All compounds described above are generally also used in order to schedule the diet and to estimate the food quality. Their concentrations can be easily found in tables of food composition of the most developed Countries.

Unfortunately, nutritional tables give us just an information on the chemical presence and quantity of these compounds in single foods, and do not give us any information about the relationships that could exist between them. The problem of this lack of information also happens when we schedule a diet. The only unifying value is the energy density, expressed in Kcal for every 100g of product.

Generally speaking, when scientific literature considers the most common macro and micro nutrients used for defining the food quality, in our opinion, it forgets or underestimates the role played by antinutrients. For example, in food quality we have to pay a particular attention on them for diet scheduling. In fact how can we talk about quality, if in a diet, together with the nutrients, are present the antinutrients, too? Which kinds of antinutrients are present and what are their amounts?

As shown in the next list", antinutrients and phytochemicals in foods could have both adverse and beneficial effects on human health. For example current recommendations suggest to consume fruits and vegetables in order to prevent diseases such as cardiovascular disorders, diabetes, ageing and cancer. On the contrary food or diets with high amount of antinutrients such as phytic acid, lectins, phenolic acids, tannins, saponins, enzyme inhibitors, cynogenic glycosides and glucosinolates, show an adverse effect on human health.

Phytic Acid

The phytic acid is a well know mineral chelating agent (zinc and calcium). Phenolic compounds can reduce the availability of zinc and during thermal processes can be combined with amino acids and make them unavailable [8].

Protease Inhibitors

Protease inhibitors are abundant in raw cereals and legumes, especially in soybeans. The antinutrient activity of protease inhibitors is associated with growth inhibition and pancreatic hypertrophy. Trypsin inhibitors give rise to inactivation and loss of trypsin in the small intestine. The amylase inhibitors give a hypoglycemic effect [9].

Saponins

Saponins are toxic compounds, because they are able to lyse erythrocytes due to their interaction with cholesterol in the erythrocyte membrane [10].

Lignans and Phenols

Lignans and phenols may cause infertility and liver diseases. However these compounds are important for cancer prevention and antioxidant capacity [11].

Lectins

Their toxicity is due to their binding with specific receptor sites on the epithelial cells of the intestinal mucosa with subsequent lesions and abnormal microvilli development. The consumption of lectins may lead to endogenous loss of nitrogen and protein utilization. Lectins may also cause systemic effects increasing protein catabolism and breakdown on steered fat and glycogen and disturbance of mineral metabolism.

Fortunately many of these antinutrients are very heat labile or can be removed by cooking processes; however in our opinion it is essential to consider also the risks for gastrointestinal functions, related to their adverse or negative action especially in population that consume, by foods, a large quantity of these compounds [7].

The nutritional values have to be preserved along all food chain processes, harvest, storage of raw products, processes, distribution and cooking processes. Undoubtedly food transformations can decrease the nutritional value of food, or can help us in order to destroy heat labile antinutrients or remove them from final products, even if in this case heating can rise problems because the thermal reaction between starch and asparagines produces acrylamide, a toxic compound [12].

NONMEASURABLE PARAMETERS

Generally speaking the quality depends not only on the choice of measurable parameters, allowing to determine food quality in an objective way, but also on non "scientific" and nonmeasurable aspects.

The consumers' choices are influenced by their own personal, ethical and emotional experiences. On these bases they interact with foods, defining the quality and giving evaluations in a wide range of terms: bad vs. good, like vs. dislike, pleasing vs. unpleasing, favourable vs. unfavourable, healthy vs. unhealthy, safe vs. unsafe etc.. All these terms rise from different psychological, ethical and religious attitudes. These social and anthropological aspects are strictly related to the consumer's behaviour on the food choice [13].

Another way of defining quality consists in describing in the Green Paper of E.U [14], in which the concept of origin and geographical area, the methods of production, the traceability and the related certifications (International Standards Organization (ISO), Protected Geographical Indications (PGI), Protected Designation of Origin (PDO) and Traditional Specialities Guaranteed (TSG)) are stressed. But also in this Green Paper we don't deal with measurable parameters but just with different classes in which single foods can be included.

The idea of quality is often confused with standardization, that means that a product has always the same taste, the same palpability, the same flour, the same colour and the same form in every Country (as for McDonald's). In this way consumers have always assurance that at every latitude they will find the same expectations. We can have products with low quality, medium quality and high quality. The important is that they are always and everywhere the same.

In our opinion another mistake, present in concepts and opinions expressed in many studies and conferences, considers the single properties of each food without any relationship with the consumers' real life. We prefer to consider food as a tool and diet as a process applying it.

In a general way useful parameters involved in quality determination are not comparable to each other (i.e. different measure units and scales). In addition, in literature several quality measurement techniques make use of weights - associated to the characteristic parameters - which introduce a sort of subjectivity in defining quality.

In the past, quality was, and still today is, in a prescience stage. Quality is not defined in a totally objective, but still in partially subjective, way. A suitable comparison could be established with Aristotelian physics. Aristotle knew that there are some relations between objects we can observe but he did not establish quantitative relations in order to describe phenomena in a correct and modern way. Archimedes first (who was "summis ingeniis dux et magister" [15], and Galileo Galilei later, changed physics and established for the first time quantitative relations between phenomena.

Archimedes was an enthusiast of mathematical method and applied it to his researches. Similarly, in order to take quality out of subjective domain and put it into an objective domain it is necessary to apply the mathematical method. It is true that in order to determine and reach high quality of a particular product it is necessary to establish the goals we want to reach. But this does not change the fact that the method to follow must be mathematically rigorous. The chosen method intends to measure several characteristics of the examined product, related to some features of quality, unbinding somehow them from their units of measurement, and finally joining them in order to give a holistic vision of the quality of the product under examination.

Paraphrasing Galileo, quality is "written in mathematical language, and its characters are formulas, equalities, inequalities and other algebraic tools, whose absence gets impossible to realize anything, without them it is like to roam vainly in a dark labyrinth".

In order to solve these methodological problems, for our purposes we developed a Functional Mathematical Index (FMI) that allows to define the quality in food and processes without weights. FMI is a number obtained by a suitable choice of a set of parameters that in a unique way characterizes the quality of food and processes. The choice of number and type of parameters allows to make some particular features to raise up.

Mathematically speaking FMI was introduced as the norm of a vector in an N-dimensional space, where N is the number of chosen parameters. The chosen parameters were normalized (their values are between 0 and 1), and their squares were summed together. Further studies of our research group have modified and improved the mathematical meaning of FMI, which however can be interpreted, in a very general way, as the distance from the "ideal" food and the considered sample. The better the quality is, the closer to 0 FMI is. The core of FMI building process is the appropriate choice of the parameters.

The FMI shows a wide ductility and adaptability to a wide range of foods and to different application fields.

In order to evaluate the importance and the usefulness of FMI, in this review we will summarize the developments of this index and we will focus on the last definition applied onto food and processes

THE MATHEMATICAL FORMULATION

The Normalized Indexes

In [16, 17] we proposed a Functional Mathematical Index (FMI), which allows the description of the total quality of an olive oil by means of a (even simple) formula, where all the parameters are comparable, but which could be modified very easily, according to different needs in order to weigh some parameters more than others. The target of the index was to establish a quantitative method to determine the "total quality" for each olive oil and detect the critical points, during all the production chain, on the basis of experimental data related to the evolution of quality indexes during the different phases of maturation grade of olives, and to show which parameter values it would be possible to modify in order to increase nutritional and safety quality or optimize the technological processes. Another important requirement was to cancel units measurement influence.

On one hand, this tool was mainly addressed to producers, to help them to detect the critical points along the whole food production chain and eventually operate the needed corrections in order to obtain a high quality product. On the other hand, the FMI could be helpful also for consumers, because usually the unique indication that they can use today, in order to operate a choice, is mainly the price of the oil, which is, in general, not necessarily related to its quality.

FMI does not pretend to be the unique answer to the need of a "quality index", but wants to show a way through which we can reliably establish if a product can be considered of high quality.

Main feature of our index is its flexibility, because it can be adapted to every choice of the quality parameters. In fact, as we will show, it is sufficient to modify the geometrical entity, related to the parameters chosen, for weighing the quality.

Let us explain in more details the concepts above exposed. If we have a pool of products and we want to determinate the "best of them", by 9 parameters (see table 1), we have to know the analytical data of each single parameter and product. We can also divide the 9 parameters into three groups, which are respectively related to nutritional, safety and technological process quality, and then develop the mathematical index in terms of them.

The three groups are:

Nutritional quality parameters: A, B, C, D
Safety parameters: E, F
Technological process parameters: G, H, I

Experimental data show that, if we are looking for high quality product, the values of each parameter must have an upper bound and a lower bound. Some parameters are chosen on the basis of international laws or recommendations, others are chosen on the basis of literature data or by some constraints. In other cases, they are simply the maximum and minimum values of all studied samples. They are all *a priori* fixed.

Let us call respectively x_n^M and x_n^m the maximum value and minimum value of each parameter range.

In Table 1 we show the ranges for all the 9 parameters; of course, in the spirit of this work, the optimal range of each parameter can be changed, depending on the different National and/or local laws.

Table 1. Parameters with related lower and upper bounds

Index	Parameters	Lower bound	Upper bound
1	A	7.5 %	20 %
2	B	0.5 %	5.0 %
3	C	55.0 %	83.0 %
4	D	3.5 %	21.0 %
5	E	0 %	0.8 %
6	F	0 meq/O_2	20 meq/O_2
7	G	-1.6 Ka/Kc	1.6 Ka/Kc
8	H	1.2 mg/100g	43 mg/100g
9	I	20 mg/Kg	900 mg/Kg

In general, recalling Table 1, we can divide the set of parameters into three groups:

1. Parameters of the first kind (or centered), whose range is symmetric with respect to the best (mean) value $\overline{x}_n = \dfrac{x_n^M + x_n^m}{2}$: the nearer to this value the parameter is, the better the quality is (Table 1, indexes 1 to 4); the choice of the mean value as the best value is rather arbitrary and can be changed appropriately;
2. Parameters of the second kind (or less): the smaller its value is, the better the quality is (Table 1, indexes 5 and 6);

3. Parameters of the third kind or "saturating parameters" (or more): the higher its value is, the better the quality is; but if the value exceeds x_n^M, then the index determines a penalization (Table 1, indexes 7 to 9).

The *centered* parameters represent those observable properties whose optimum is the average of the two extreme values. The *more* parameters represent those observable properties whose optimum is the maximum value. The *less* parameters represent those observable properties whose optimum is the minimum value.

The parameters must be normalized, in order to obtain, for every index, values varying in the interval [-1, 1].

For the first group the index is defined in the following way:

$X_n = \left(\frac{x_n - \overline{x_n}}{r_n} \right)$, where represents the radius of the interval, that is its semi-length; for the second group we consider ; for the third group we put

$$X_n = H(x_n - x_n^M) + sign(x_n - x_n^M) \cdot \frac{x_n - x_n^M}{x_n^M - x_n^m} =$$

$$\begin{cases} \dfrac{x_n^M - x_n}{x_n^M - x_n^m} & if \quad x_n \leq x_n^M \\[2ex] \dfrac{x_n - x_n^M}{x_n^M - x_n^m} + 1 = \dfrac{x_n - x_n^m}{x_n^M - x_n^m} & if \quad x_n > x_n^M \end{cases} \tag{1}$$

where $sign(t) = \begin{cases} 1 & if \quad t > 0 \\ 0 & if \quad t = 0 \\ -1 & if \quad t > 0 \end{cases}$ and $H(t) = \begin{cases} 1 & if \quad t > 0 \\ 0 & if \quad t \leq 0 \end{cases}$

is obtained from the so-called Heaviside step function [18], which allows the index to jump discontinuously from 0 to 1, when the parameter exceeds its maximum allowable value. Differently from the Heaviside function, we put $H(0)=0$, in order to guarantee that, when $x_n = x_n^M$, the corresponding index becomes 0.

Penalizations and Generalizations

Formula (1) can be generalized, depending on how much the product or the process under examination, whose parameter values do not belong to the allowed intervals, must be penalized. In this case we can vary the slope of the line for and $x_n > x_n^M$:

$$X^{\alpha}{}_{n} = \begin{cases} 1-\alpha(x_n - x_n^m) & if \quad x_n < x_n^m \\ \dfrac{x_n^M - x_n}{x_n^M - x_n^m} & if \quad x_n^m \le x_n \le x_n^M \\ 1+\alpha(x_n - x_n^M) & if \quad x_n^M < x \end{cases} \tag{2}$$

The greater the value of $\alpha$is, for and $x_n > x_n^M$, the steeper the line is. In other terms, α can be interpreted as a penalization coefficient.

Formula (1) can be obtained from (2), putting for and $x_n > x_n^M$.

Let us observe that, for every value of α, when x_n tends to from above, and when $x_n$tends to x_n^M from below, tends to 1; furthermore, when x_n tends to x_n^M from below, tends to 0. In mathematical symbols

$$\lim_{x_n \to (x_n^{MAX})^+} X^{\alpha}{}_n = 1 \ ; \quad \lim_{x_n \to (x_n^{MAX})^-} X^{\alpha}{}_n = 0 .$$

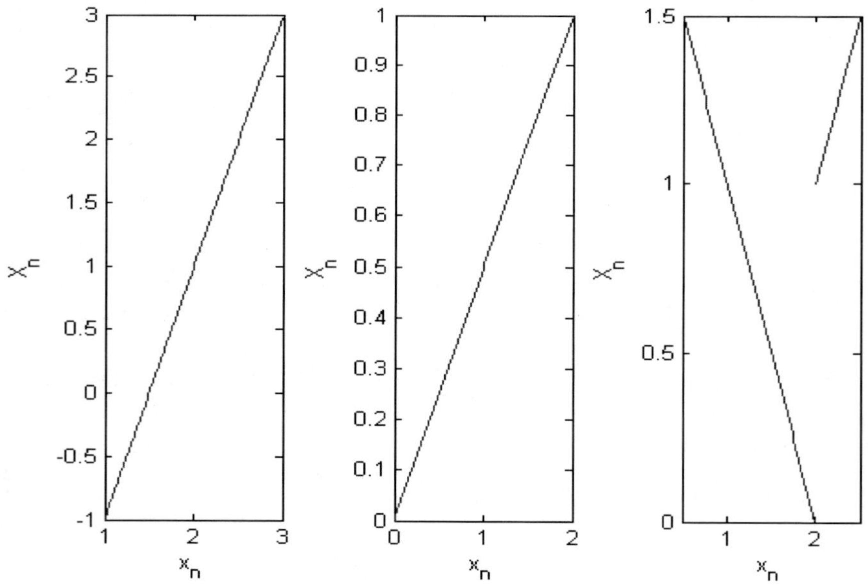

(a) (b) (c)

(a) first group (parameters 1 to 4 in Table 1); (b) second group (parameters 5 and 6 in Table 1); (c) third group (parameters 7 to 9 in Table 1).

Figure 1. Qualitative behaviour of the indexes for the three parameter groups.

In (a) and (c) we have chosen $x_n^m = 1$; $x_n^M = 2$; in (b) we have chosen $x_n^m = 0$; $x_n^M = 2$.

When $x_n^m \le x_n \le x_n^M$, the two indexes X_n are defined as in Figure 1; for $x_n > x_n^M$ and $x_n < x_n^m$, we have arbitrarily put $X_n = 1.5$.

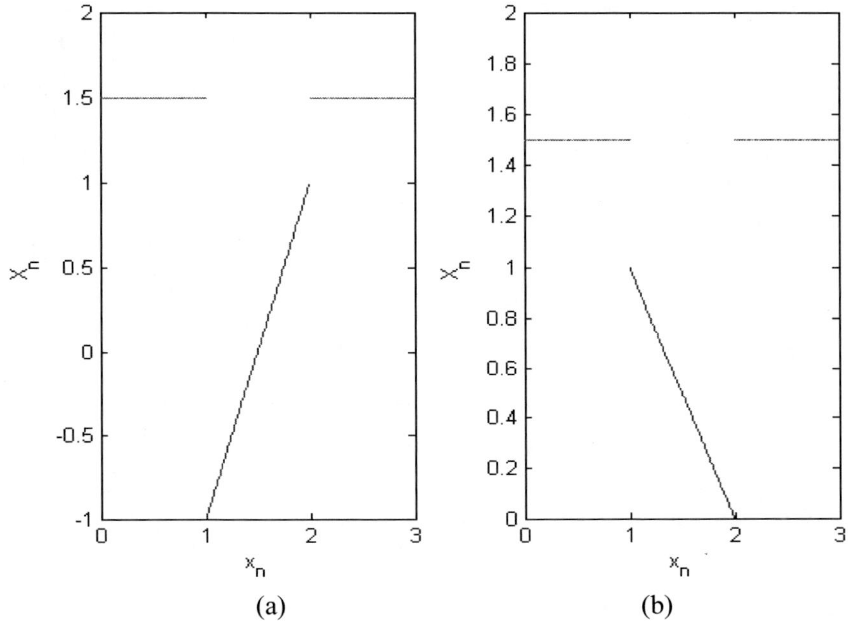

Figure 2. Modified indexes for the first and third parameter group, with a constant penalization function.

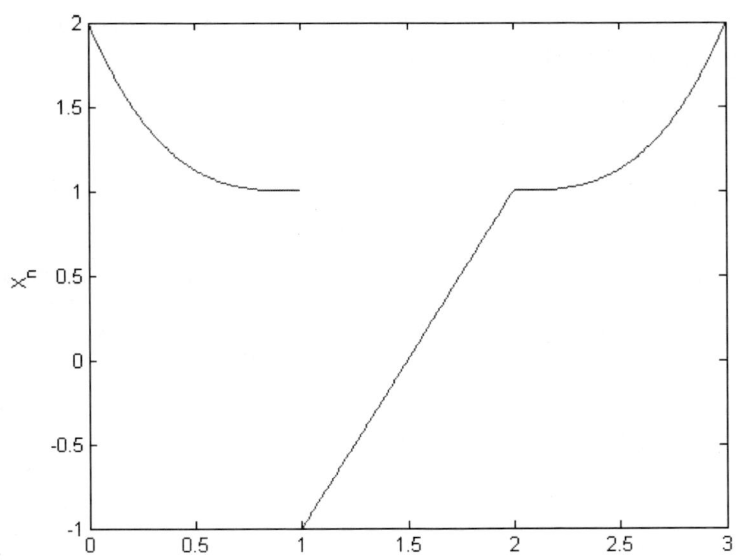

Figure 3. Modified index for the first parameter group, with a polynomial penalization function.

When $x_n^m \le x_n \le x_n^M$, the indexX_n is defined as in Figure 1. We have chosen $x_n^m = 1$; $x_n^M = 2$, $N = 3$.

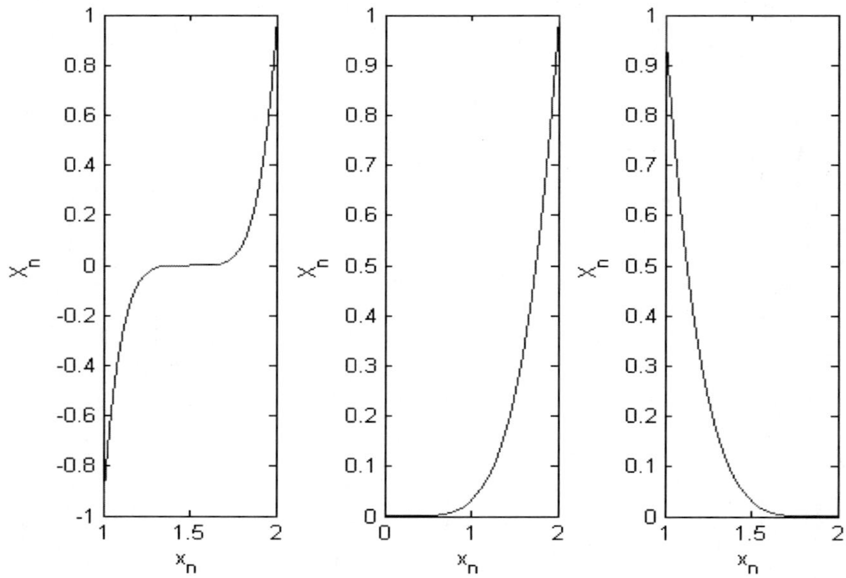

(a) (b) (c)

In (a) and (c) we have chosen $x_n^m = 1$; $x_n^M = 2$; in (b) we have chosen $x_n^m = 0$; $x_n^M = 2$. $N = 5$.

Figure 4. Indexes of polynomial type for $x_n^m \le x_n \le x_n^M$.

In Figure 1 we have plotted the qualitative behaviour of the three parameter groups. The indexes related to the first and third group can assume values not belonging to the interval [0,1]. For the second group the situation is completely different, because for the parameters E and F upper bounds (respectively E 0.8% and F 20 meq/O_2) cannot be exceeded, (i.e. due to International or National regulations, literature etc.

FMI as a Distance

Let us introduce the Euclidean space $\mathbf{R}^N$ (where N is the number of parameters studied), endowed with the usual Euclidean metric or norm: $\|x\|_1 = \sqrt{\sum_{i=1}^{N} x_i^2}$ for every vector x = [19].

The Euclidean norm of the vector $\boldsymbol{X}$, whose components are the indexes X_n, represents our "Global or Total Quality" index (or Global or Total FMI)I_{GQ}:

$$I_{GQ} = \|X\|_1 = \sqrt{\sum_{n=1}^{N} (X_n)^2}$$

With our choice, the optimal product corresponds to the vector X = (0,0,...,0), that is the origin. The worse the quality is, the higher $||X||_1$is. In fact, every component of the vector represents the normalized distance of the parameter from the optimal value. Thus a high norm means a poor global quality; a good product has a small value of I_{GQ}; a poor product has a high value of I_{GQ}.

To a parameter whose value belongs to the allowed quality range corresponds an index such that $|X_n| \leq 1$. Consequently, to a food with only good values of the quality parameters corresponds a vector, in the *N*-dimensional space, whose norm is less or equal to $\sqrt{N}$. In other words, if a product has a high global quality, its quality vector must belong to the *N*-dimensional hypersphere S_N, centered in the origin and with radius of length $\sqrt{N}$. The farther from the origin the vector, the worse its "global quality". The nearer the origin the vector, the better its "global quality".

The index I_{GQ}can be interpreted as a "global" index, because its value is the sum of *N* contributions by every single component. However, belonging to the *N*-dimensional hypersphere S_N does not guarantee a high quality. In fact, if it is true that, for every oil whose parameters belong to the intervals shown in Table 1, the corresponding index vector X belongs to the hypersphere S_N, the reverse implication, in general, is no longer true, except for the indexes 5 to 6 (see the above discussion). If, for example, a food is characterized by the best values of the parameters, apart from one, to which corresponds a "low quality" index (for example $X_n = \sqrt{N}$), much greater than 1, nevertheless the vector X belongs to the hypersphere; in fact, the vector X = (0,0,...,0, $\sqrt{N}$,0, ..., 0) is such that $||X||_1 = \sqrt{N}$. In other words, belonging to the *N*-dimensional hypersphere $S_N$is only a *necessary* condition for a high quality product.

In order to consider (and correct) similar situations, we introduced a second index, that represents, in some sense, the "Local Quality" index (or Local FMI) I_{LQ}:

$$I_{LQ} = ||X||_2 := \max | X_n | \; ; n = 1, \ldots, N$$

The set of the vectors verifying the condition $I_{LQ} \leq 1$ is the *N*-dimensional hypercube C_N, centered in the origin and with length of its edge equal to 2, i.e., with semidiagonal of length $\sqrt{N}$. An oil whose parameters belong to the intervals shown in Table 1 is such that $I_{LQ} \leq 1$, i.e., the corresponding vector X belongs to the *N*-dimensional hypercube C_N. This hypercube is inscribed in the hypersphere S_N. Thus every vector X belonging to the *N*-dimensional hypercube C_N belongs to the *N*-dimensional hypersphere S_N.

For the sake of simplicity, let us suppose that max | X_n | = | X_1 |. We have

$$||X||_2 = \max |X_n| = |X_1| = \sqrt{X_1^2} \leq \sqrt{\sum X_n^2} = ||X||_1 \leq \sqrt{N(X_1)^2} = \sqrt{N}\,|X_1| = \sqrt{N}\,||X||_2.$$

The condition $I_{LQ} \leq 1$ excludes oils with even only one index greater than 1. Thus it is obviously a sufficient condition for high quality, because guarantees that all the parameters belong to the intervals shown in Table 1.

The reason why we adopt two quality indexes is that, while$I_{LQ} \leq 1$guarantees that no parameters exceed the values shown in Table 1 (without considering the global contribution of the parameters), I_{GQ}(representing the distance from the optimal point, i.e., the origin) describes a global property of the product. These considerations can be extended to a very wide class of products and processes.

The second index can be considered a "local" index, because gives only the information about the maximal value of the components. Consequently, the vectors X=(1, 1, ... , 1, 1) and X=(1, 0, 0, ..., 0, 0) have the same I_{LQ}. Thus we cannot neglect the global index, which tells us, for every vector whose I_{LQ} is less than 1, how much it is close to the origin, i.e., how much good it is.

Now we can consider some critical examples:

1. to X = (0, 0, ..., 0, 1, 0, …, 0)we associate $I_{GQ} = 1$; $I_{LQ} = 1$;
2. to X = (0,0,...,0,1,1,0, …, 0) we associate $I_{GQ} = \sqrt{2} > 1$; $I_{LQ} = 1$;
3. to X = (1,1,...,1,1) we associate $I_{GQ} = \sqrt{N} > 1$; $I_{LQ} = 1$;
4. to X = (0,0,...,0, $\sqrt{N}$,0, …, 0)we associate $I_{GQ} = \sqrt{N} > 1$; $I_{LQ} = \sqrt{N} > 1$;
5. to X = (0, 0, ..., $1/\sqrt{2}$, $1/\sqrt{2}$, 0, …, 0)we associate $I_{GQ} = 1$; $I_{LQ} = 1/\sqrt{2} < 1$;

In the above examples, the first, the second and the third vector have the same Local FMI, but the Global FMI of the first is better than the second and much better than the third. The third and the fourth vector have the same Global FMI, but only the third has a good Local FMI. The first and the fifth vector have the same Global FMI, but the fifth has a much better Local FMI.

In other words the Local FMI and Global FMI can compensate and complement each other.

It is a matter of discussion if and how the two quality indexes can be compared, in particular when any classification, comparison or score of different products is needed.

Generalizations of FMI

We could also operate a linear combination of the indexes X_n (where we could weigh some of them more than the other ones); or we could first require that one of the indexes must stay below an *a priori* determined threshold and only secondly, if the first condition is satisfied, we could require that the other indexes must stay below specific thresholds.

Our FMIs are flexible to any law changes or new local or international regulations, new literature data and so on. On the other hand, the previous formulas for the indexes X_n can be easily adapted to several, different, requirements. If, for example, we decide that a product exceeding the threshold for one or more parameters must be considered of poor quality, independently from how much the threshold has been exceeded, then we can decide to modify the corresponding indexes in such a way that they heavily affect the value of the total indexes.

One of the simplest ways to yield this result is to modify the indexes putting $X_n = M$ if x_n does not belong to the allowed range, where M is an arbitrary value, which can be established depending on the penalization level chosen. As observed above, the indexes of the second group cannot exceed value 1. The graphs of the two modified classes of indexes are shown in Figure 2.

If we want to save oils which slightly exceed the threshold and simultaneously heavily penalise those ones whose excess is high, we can modify the formulas introducing a "penalization – awarding" coefficient, choosing, if $x_n$does not belong to the allowed range, instead of $X_n = M$, the relation$X_n = f(x_n)$, where $f(x_n)$ is a function assuming values close to 1 when x_n slightly exceeds the allowed bounds and growing fast when $x_n$goes far away from the bounds. A polynomial $P_N(x_n) = \sum_{k=0}^{N} a_k x^k$ of degree Nor an exponential function would perfectly match these requirements. Concerning the polynomials, the higher the degree is, the greater the number of parameters a_k involved is. Actually, it would be sufficient to consider, for $x_n \geq x_n^M$, the polynomial or $P_N(x_n) = 1 + \left(x_n - x_n^M\right)^N$, which is equal to 1 when and grows with x_n and, for $x_n \leq x_n^m$, the polynomial $P_N(x_n) = 1 + \left(\left|x_n - x_n^m\right|\right)^N$, which is equal to 1 when and grows when grows (see Figure 3).

Buffer Effect in Data Processing

We can also reproduce the realistic situation of products assuming parameter values sufficiently close to the optimal ones to be in any case considered of the highest quality. In fact, on one hand it is unrealistic to expect that the products succeed in reaching some optimal values, like, for example, E equal to 0 (in this case, we can require that the product must have values sufficiently close to the optimal ones), on the other hand, as already remarked above, the choice of the mean value as the best value in the first group of parameters is arbitrary and cannot penalise products whose values do not coincide with the mean but are sufficiently close to it. As shown in Figure 4, we can take into account these requirements and minimize (buffer) the effects of each index, using polynomials of degree N greater than 1, for $x_n^m \leq x_n \leq x_n^M$. In this case, we build the following indexes:

for the first group: $X_n = \left(\dfrac{x_n - \overline{x_n}}{r_n}\right)^N$, where and $r_n = \dfrac{x_n^M - x_n^m}{2}$;

for the second group: $X_n = \left(\dfrac{x_n}{x_n^M}\right)^N$;

for the third group: $X_n = \left(\dfrac{x_n^M - x_n}{x_n^M - x_n^m}\right)^N$.

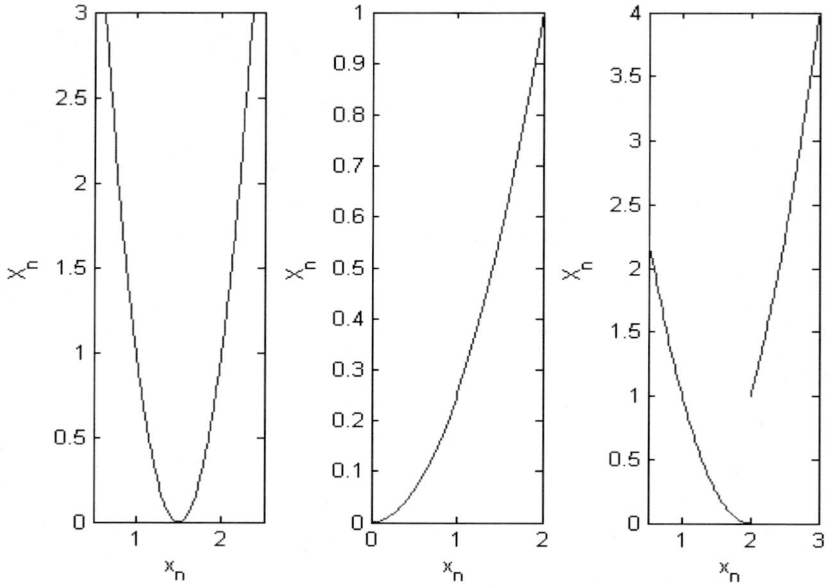

(a) (b) (c)

Modified indexes, according to the formulas chosen for the computations in [16], [17]. In (a) and (c) we have chosen $x_n^m = 1$; $x_n^M = 2$; in (b) we have chosen $x_n^m = 0$; $x_n^M = 2$. $N = 2$.

Figure 5. Indexes of polynomial type.

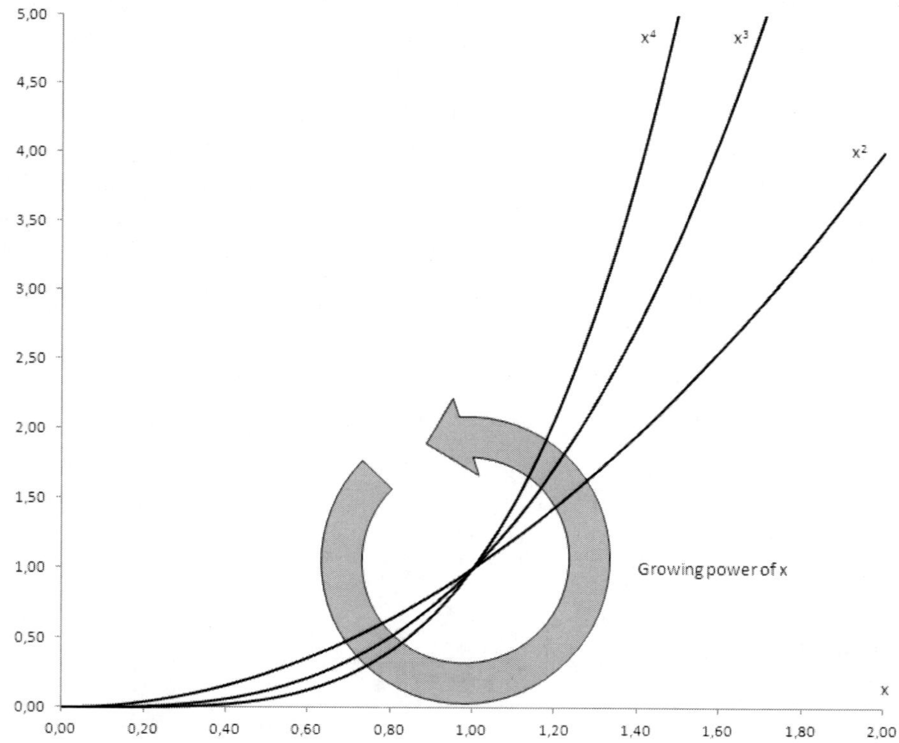

Figure 6. Buffering effect of increasing x powers.

In this way, when the X_n value are well below 1, there is a remarkable buffering effect: the higher the power of X_n in the definition of FMI, more evident is the buffering effect, i.e. X_n values near to zero become very small (they can become practically zero). This effect is not evident for X_n near the upper bound ($X_n = 1$). In this way, the high power penalizes the "bad" X_n values, which become relevant in the FMI sum, and awards the small values (they become irrelevant in the FMI sum). In Figure 6 we graphically show this buffering effect: below $X_n = 1$ the growing power flattens the curve, and this effect is more evident for values far from 1; while for values above $X_n = 1$ the power makes the curve steeper, enhancing the penalization effect. In the figure, the x-axis represents the X_n value and y-axis represents the related X_n power.

The lower bound of the Global FMI, which obviously is equal to 0, is referred to as *standard reference*, and represents the optimal (i.e., "ideal") value for a perfect product.

In order to test our Global and Local FMIs, in [16,17] we chose the following indexes (putting N=2 in the three previous formulas):

for the first group (indexes 1 to 4): $X_n = \left(\frac{x_n - \overline{x_n}}{r_n} \right)^2$, where and $r_n = \frac{x_n^M - x_n^m}{2}$;

for the second group (indexes 5 and 6): ;

for the third group (indexes 7 to 9): $X_n = \begin{cases} \left(\frac{x_n^M - x_n}{x_n^M - x_n^m} \right)^2 & if \quad x_n \le x_n^M \\ \left(\frac{x_n - x_n^m}{x_n^M - x_n^m} \right)^2 & if \quad x_n > x_n^M \end{cases}$,

with which the Global and Local FMIs were computed (see Figure 5).

The described approach makes it possible to optimize and predict the beneficial effects of different components and processes in different food categories.

In another study [20] the FMI was used to describe the derivation and application of a new functional mathematical index that was used to evaluate the nutritional, safety, and processing quality aspects of potatoes. The index introduces the concept of an "optimal potato", to establish whether a specific potato variety or processed potato product can be considered of high, medium, or low nutritional quality. The main goal of the paper was to link composition and chemical quality with factors in growth, production, distribution, and processing of potatoes and potato products for commercial use.

In several studies not all the indexes used have the same weight (each index contributes in a different way, and it is very difficult attribute to each of them the correct role played in this system). Therefore, we can, on one hand, introduce the following new weighted formula:

$$I_{GQ} = \sqrt{\sum_{n=1}^{N} w_n \cdot (X_n)^2} \quad \text{where } w_n \text{ are the weight and } \sum_{n=1}^{N} w_n = 1$$

In general, this is no more a distance in a mathematical sense [21]; however, it clearly preserves the main properties of a distance in a practical sense: the farther from the origin the vector, the worse its "global quality". The nearer the origin the vector, the better its "global quality".

In the paper we use the fourth power of X_n instead of the second power as above:

$$I_{GQ} = \sqrt{\sum_{n=1}^{N} (X_n)^4}$$

We also introduced [20] a new set of parameters that are more suitable to build the FMI describing the quality aspects of potatoes.

In [22, 23, 24] the mathematical aspects of the FMI were refined and applied to derive relationships to the reported wide-ranging composition of phenolic compounds (flavonoids) of eight sweet potato cultivars. In [24] the index was defined by the following formula:

$$I_{GQ} = \sqrt{\frac{\sum_{i=1}^{n} (X_i)^4}{n}}.$$

considering that, since the sum of n good quality parameters can have a maximum value of n, it must be divided by n, the number of parameters being evaluated. In this way also the FMI of good quality products is normalized.

In a further study [25] we used the generalization of the FMI, applying it to different features of foods (not only the quality). If the product has different (independent) features (for instance, dual biological activities, as antibacterial and antitumor properties in the case of particular molecules [25]), independent definitions for partial FMIs (named subFMIs) can be produced using only some of the parameters. In this case the normalization with respect to the number of parameters characterizing each subFMI becomes essential, in order to compare the contribution of every subFMI.

Just for the sake of clarity, in [22] and in further papers, the indexes X_i were defined "local FMIs", in the sense that while some parameters can be grouped to build a subFMI, some others can contribute by itself to the definition of some other, peculiar, properties which cannot be grouped. Thus, in this case every single index can represent a local FMI, in this new acceptation. The symbol loc_i was also used, instead of X_i. This symbol will be used in the FMI Workbench Section.

From a mathematical point of view, it is possible to nest an FMI inside another (a more general or "global") FMI. Indeed, it is possible to consider a set of subFMIs and embed them into a single parameter raised to the second power. The previous FMIs then become new local FMIs (the above subFMIs) of an extended FMI obtained considering all the product features. For this application, we obtained the following general definition:

$$FMI = \sqrt{\frac{\sum_{i=1}^{n_1} X_i^4}{n_1}} + \sqrt{\frac{\sum_{i=1}^{n_2} Y_i^4}{n_2}}.$$

where X_i and Y_i refer to the local indexes belonging respectively to the first and to the second subFMI. The normalization with respect to the number of indexes belonging to each subFMI is needed in order to compare the contributions of the two subFMIs.

In [26] a more refined formula was used:

$$FMI = \sqrt{\frac{\sum_{i=1}^{n_1} loc_i^4 + \sum_{j=1}^{n_2} subFMI_j^2}{n_1 + n_2}}$$

In this definition, two different types of indexes were considered; the first sum refers to single parameters that are not included in any subFMI, the second sum is extended to all subFMIs. Since the FMI definition can be recursive, a subFMI can in turn always be defined as a nested FMI.

As an application of the previous formulas, in [25] we extended the development of the FMI concept to define an "optimum tea" based on reported relationships between the content of structurally different polyphenolic compounds called catechins of a large number of teas and their dual beneficial effects: antimicrobial activities against a foodborne pathogen and inhibition of human cancer cell lines.

In fact, tea leaves produce secondary metabolites that are involved in the defense of the plants against invading pathogens. In the case of green teas, these metabolites are the catechins.

Because we previously reported that individual polyphenolics tea ingredients and freshly brewed teas exhibited anticarcinogenic and antimicrobial properties, the main objective of this study was to define mathematical indexes for these two beneficial effects of teas in terms of their known catechins content.

The FMI was also employed in order to classify potable waters [27], and to determine which potato cultivars are more suitable for thermal processes (baking, cooking, frying at high temperatures), guaranteeing at the same time low amounts of acrylamide and high quality levels [28].

In order to simplify the concepts discussed in this review, in the next subsection we report an numerical example of FMI calculation.

Numerical Example

To get an insight in FMI calculations, in this subsection we give an example of complete evaluation of the FMI formula for a hypothetical food. The set of data is listed in the following Table:

Table 2. Numerical example

Experimental value	Optimal value	Minimum value	Maximum value	Parameter kind
10.0	0.0	0.0	15	less
3.0	0.0	0.0	7	less
0.1	0.7	0.0	0.7	more
0.3	1.3	0.0	1.3	more
1.2	1.8	1.2	2.4	centered
9.0	8.0	6.0	10.0	centered

In order to compute the FMI, we compare the original formula, given by the Euclidean norm, and the formula with the fourth powers. In this example, in order to appreciate the buffering effect, we do not normalise the FMI, i.e., we do not divide the sum by n.

Since we have six parameters (two for each parameter kind), we can call them $X_1, \ldots, X_6$:

1. first kind parameters (centered) X_5 and X_6: $X_5 = \frac{1.2 - 1.8}{(2.4 - 1.2)/2} = -\frac{0.6}{0.6} = -1$,

$X_6 = \frac{9.0 - 8.0}{(10.0 - 6.0)/2} = \frac{1.0}{2.0} = 0.5$;

2. second kind parameters (less) X_1 and X_2: $X_1 = \frac{10.0 - 0.0}{15.0 - 0.0} = \frac{2}{3} = 0.67$,

$X_2 = \frac{3.0 - 0.0}{7.0 - 0.0} = \frac{3}{7} = 0.43$;

3. third kind parameters (more) X_3 and X_4: $X_3 = \frac{0.7 - 0.1}{0.7 - 0.0} = \frac{6}{7} = 0.85$,

$X_4 = \frac{1.3 - 0.3}{1.3 - 0.0} = \frac{10}{13} = 0.77$.

Applying the definition of FMI as a norm (sum of squares), we obtain:

$$FMI = \sqrt{X_1^2 + X_2^2 + X_3^2 + X_4^2 + X_5^2 + X_6^2} = \sqrt{0.67^2 + 0.43^2 + 0.85^2 + 0.77^2 + (-1)^2 + 0.5^2} =$$
$$= \sqrt{0.45 + 0.18 + 0.72 + 0.59 + 1.00 + 0.25} = \sqrt{3.19} = 1.79,$$

At the end each parameter for each sample can be tabulate as represented in the table 3

Using the *new* FMI (i.e. with the fourth power of X_n) the results will appear as in the following formula:

$$FMI = \sqrt{X_1^4 + X_2^4 + X_3^4 + X_4^4 + X_5^4 + X_6^4} = \sqrt{0.67^4 + 0.43^4 + 0.85^4 + 0.77^4 + (-1)^4 + 0.5^4} =$$
$$= \sqrt{0.20 + 0.03 + 0.52 + 0.35 + 1.00 + 0.06} = \sqrt{2.17} = 1.47$$

Table 3. Local and Global FMIs

Sample	X_1^2	X_2^2	X_3^2	X_4^2	X_5^2	X_6^2	Global FMI
A	0.45	0.18	0.72	0.59	1	0.25	1.79

Table 4. Local and Global new FMIs

Sample	X_1^4	X_2^4	X_3^4	X_4^4	X_5^4	X_6^4	Global FMI
A	0.20	0.03	0.52	0.35	1	0.06	1.47

These values must be compared with the maximum value for both the definitions. Referred to the original definition of FMI, we can safely say that this product has a very low quality index, because the FMI value is proximal to the maximum allowed value. Note that the X_5 parameter is borderline. On the other hand, using the new definition we can appreciate the "buffering" effect of the fourth power; in fact, the Global FMI value has been lowered to a "more acceptable" value of 1.47. But the boundary value of X_5 penalizes the total score and the lowering effect is not so evident as in the case in which all parameters are far from the boundary value.

Similar calculations can be performed using an arbitrary n-th power formulation, as shown before. The usual second power must be substituted by the corresponding n-th power of each index X_n defined above.

FMI WORKBENCH

Why a Software

In this section we present a software that is able to perform the FMI score especially when several samples with several parameter have to be processed.

The FMI definition uses several mathematical operators. So its calculus can be rather difficult and error-prone. In addition in its applications FMI must be calculated for dozens of samples. As a consequence in past necessity to develop a software able to make easy calculating this quality index rose up.

FMI Workbench (http://sourceforge.net/projects/fmiworkbench) is a software application whose main aim is to make easy FMI calculation, over an ensemble of samples organized into groups belonging to a project (see next Subsections). It is possible to import projects already written or to export them and to execute the main operations over data to be managed and to show and modify samples and groups, to add or delete data and so on [29].

UML Class Diagram

The following UML class diagram [30] (Figure 7) describes how object classes are related to each other into FMI Workbench. This is a simplified schema showing only main relations. **SubFMI**, a special case of component, is not reported.

Projects

The project is the main entity of FMI Workbench. A project states a relation between components, groups and samples on which we calculate FMI. Every project has some properties: name, description, author (or more authors) and a revision number. The project schema is made by an ensemble of components, each one representing a local FMI. The definition of the components characterizes the way FMI is calculated on every sample. Samples are organized into groups. In Figure 8 there is a representation of a project.

Components

The components of a project (Figure 9) represent the parameters with which we can calculate FMI. They have the following properties: name, unique for a component in a project, description, type, that represents the normalization strategy of the parameter, and extreme values (**High value** and **Low value**) with which we calculate the normalization. A special component, the subFMI component, represents the FMI of another project. In order to give sense to the calculus of this subFMI for a particular sample, it is necessary that another sample with the same name in the same group of the original one must belong to the project w

Groups

A group is simply an object to which samples belong. A group belongs to a project. It is used in order to group samples logically related. It is necessary to define at least one group in order to assign samples to a project. In order to define a group it is necessary to assign a unique name inside of the project and eventually a description. In Figure 10we list two example groups of a project.

here the subFMI is defined.

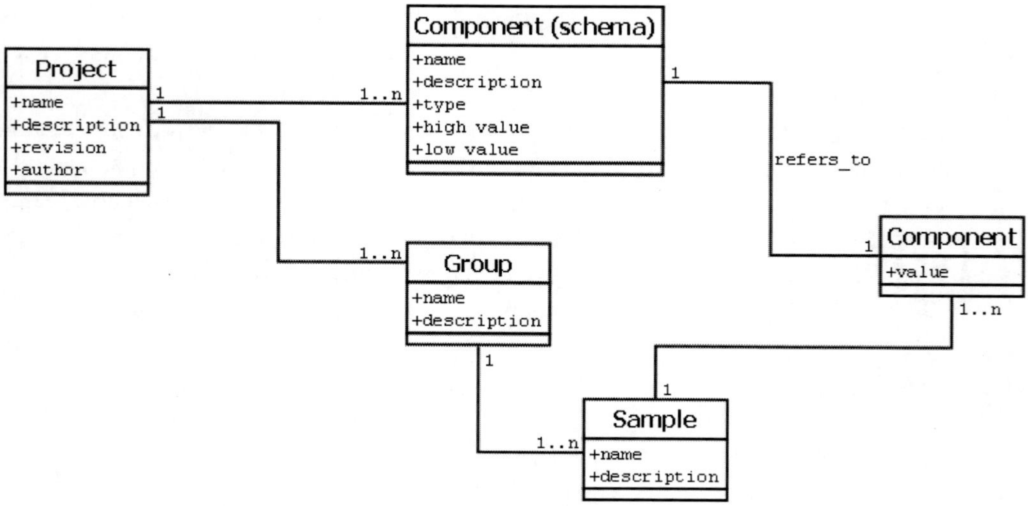

Figure 7.

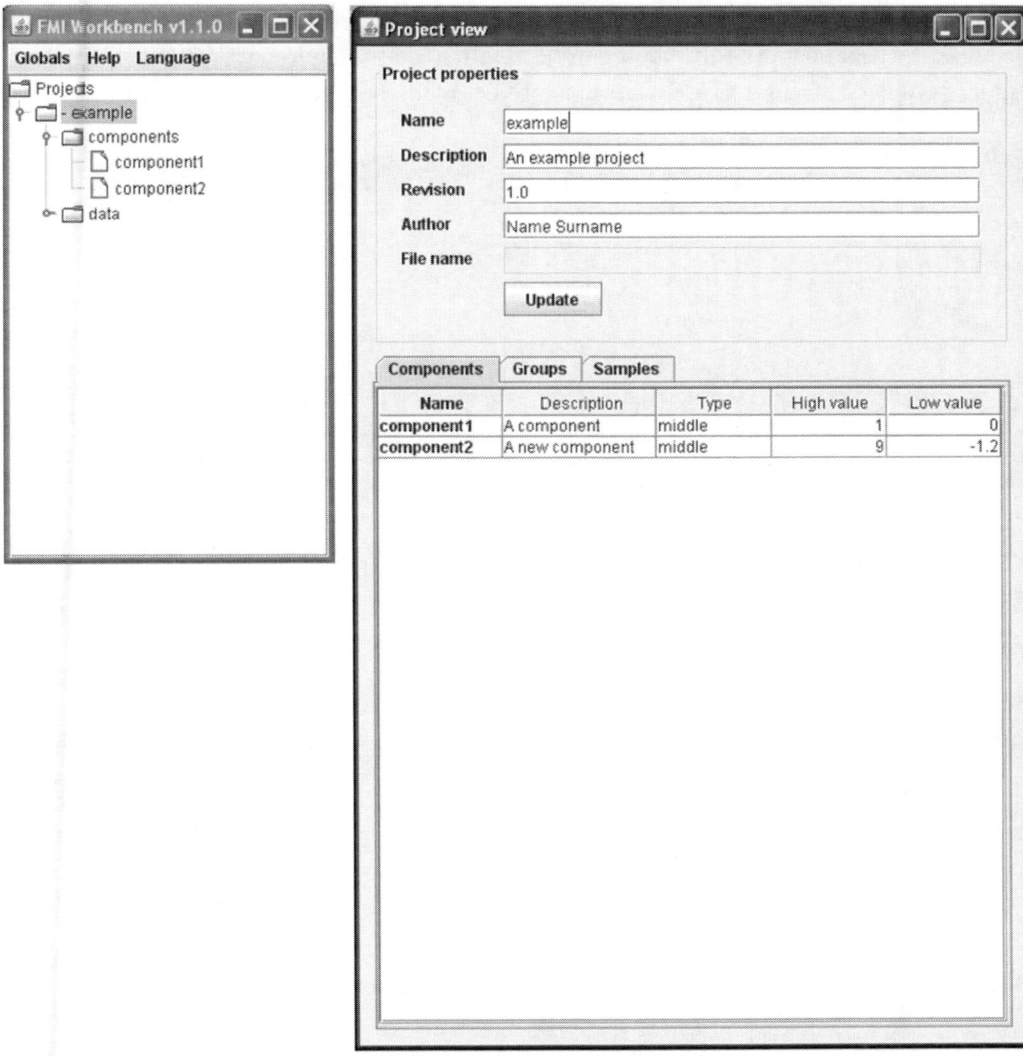

Figure 8.

Samples

A sample represents the object on which we calculate FMI. All samples have some properties: name, unique in the group, and description. For every sample it is necessary to define values for all parameters in order to calculate its FMI. In Figure 11we show a representation.

Dual Project

Project data, i.e. samples, are organized into groups. If we want to exchange groups with samples, we can create a dual project. For example, let us suppose that we have a project whose samples are organized in yearly groups. Every group, i.e. every year, will have samples of that year. But if we want to have samples organized by type, i.e. a group for every type of sample, how could we do? We can create the dual project of the initial project. A dual project has the same name of the source project with added suffix (dual). So, for example, if the source project has name: project, the dual project will have name: **project (dual)**.

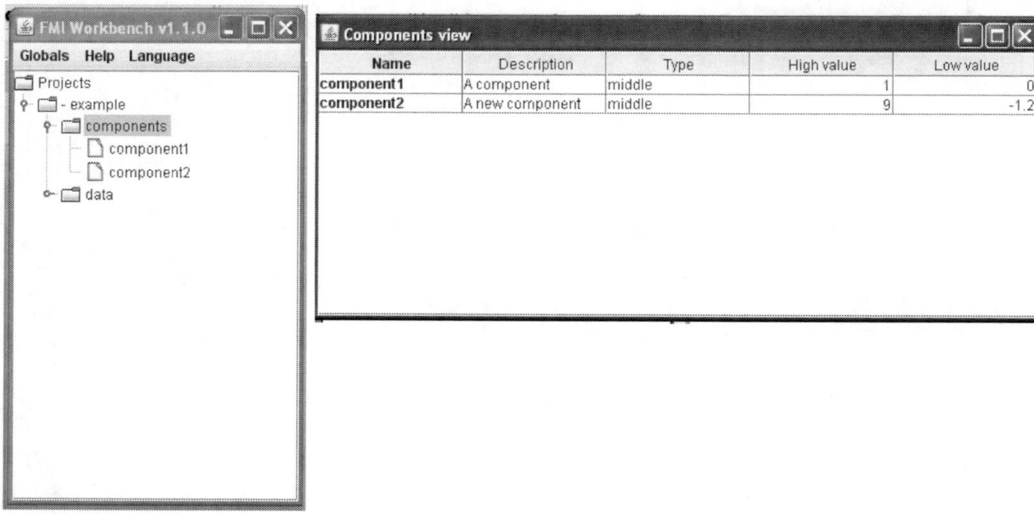

Figure 9.

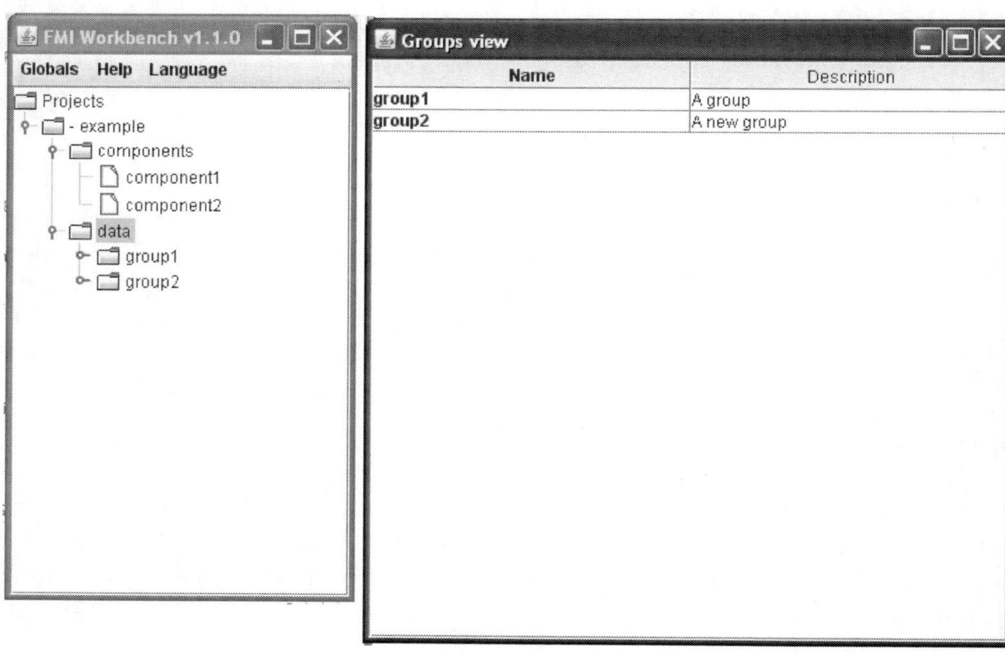

Figure 10.

In Figure12 we have represented a project and its dual. As one can notice, in **example (dual)** groups and samples are exchanged. Components and their types are not modified.

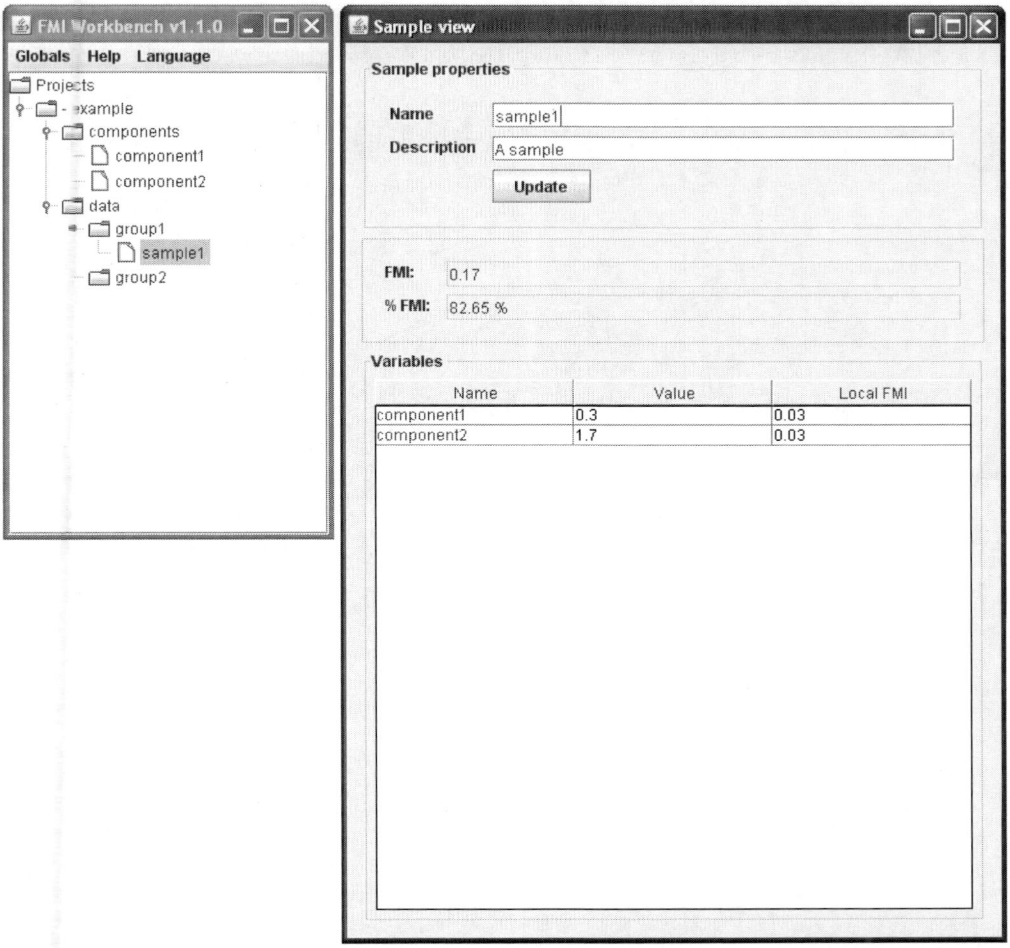

Figure 11.

Nested FMI (subFMI Components)

When we want to use a nested FMI, we have to create a new component referring to a nested project. So at first we create the nested project with its groups and its samples, and later we create the nesting project. Of course the two projects must have the same groups and the same samples, otherwise we cannot associate correctly the nested FMI to samples in order to calculate the total FMI. In Figure 13 the creation of a subFMI component is showed.

As one can see the two projects, **example** and **example_simple**, share the same groups and the same samples.

As can be noticed in Figure 14, the component **subc** has no high and low values. For a subFMI components these properties have no meaning, because they depend on the nested project and its samples.

In Figure 15 we see a report of every sample of the project **example**, with experimental values. In Figure 16 we represent local FMI values obtained by means of the previously inserted experimental values (just after having simply clicked on "Local FMI" checkbox). As can be noticed, **subc** component has no real value associated, because its value has meaning just as local FMI value, as we can see in Figure 16.

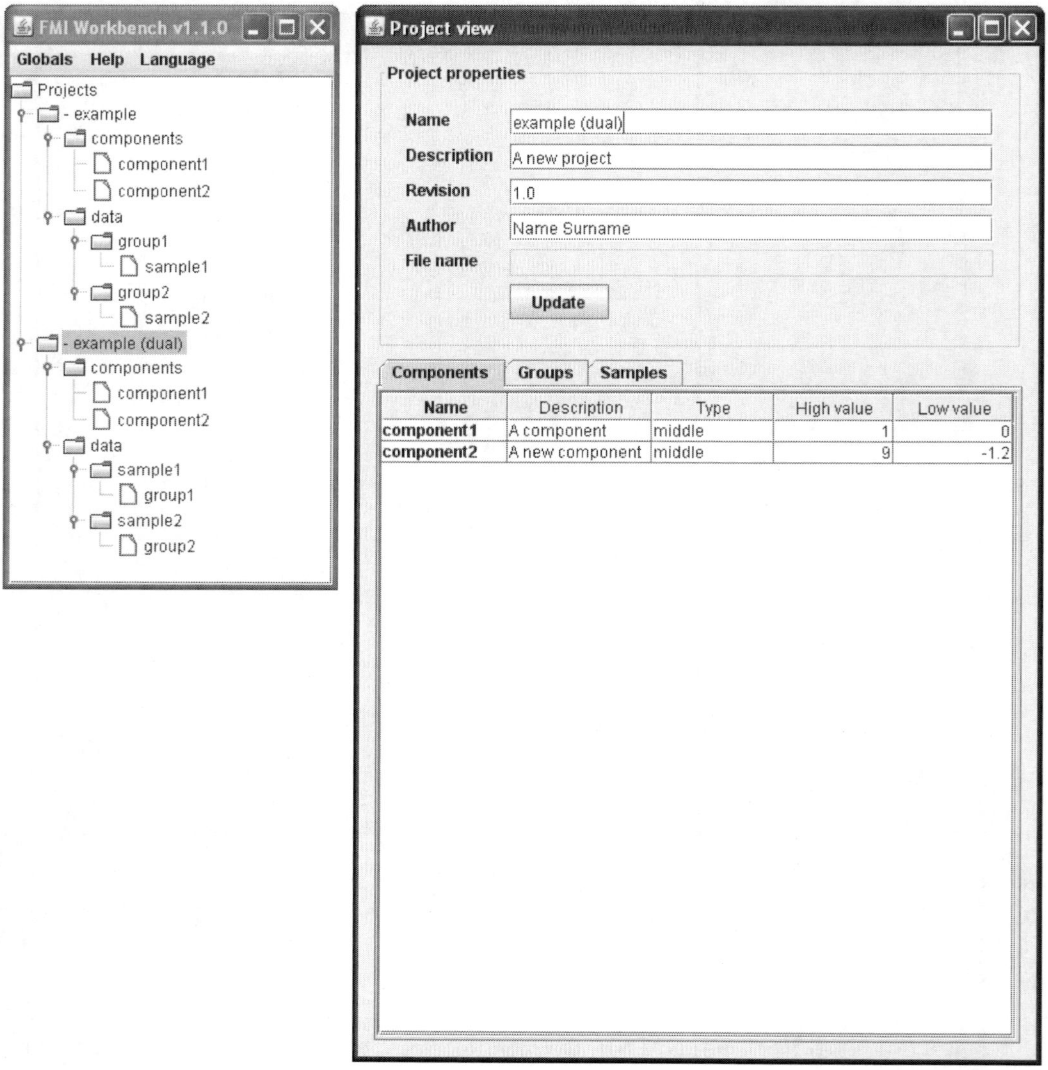

Figure 12.

Loading and Saving Projects

Of course, FMI Workbench allows to open and save project files. Its native format is XML. Projects can also be saved (and opened) in Microsoft Excel file format but some properties are lost, like Description, Author and Revision.

Microsoft Excel File Format

Excel files must follow a special format in order to be correctly interpreted by FMI Workbench. First of all a sheet named **Schema** must be present. In this sheet the project schema is described. As reported in Figure 17, the format is simple and intuitive.

For every group a special sheet must be present. The name of the sheet is the name of the group. An example is shown in Figure 18. Every sample is described by its own properties. The FMI field is used just when the file is written, and anyway it must be present. The special sheet named "-" is a collection of all samples of all groups (Figure 19).

In addition FMI Workbench can generate PDF reports of the projects. In these reports, all information related to the project appears.

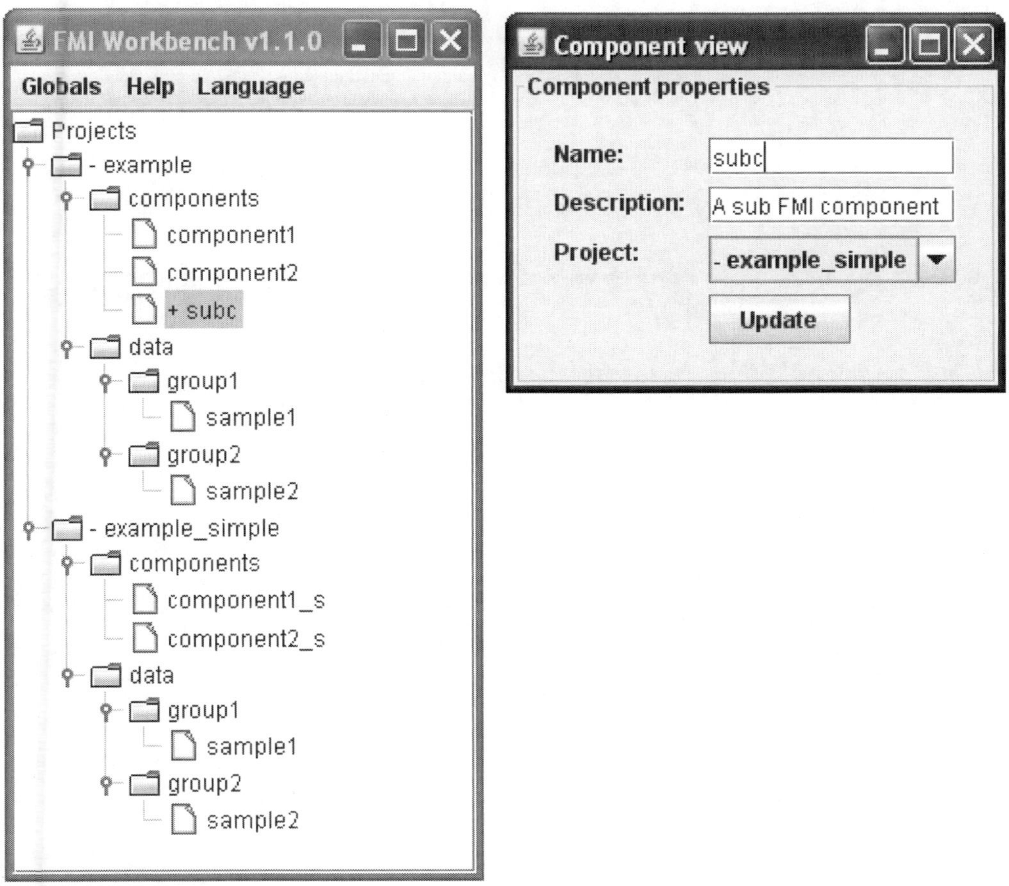

Figure 13.

CONCLUSION

The FMI index provides useful indications about the global quality of a specific product. When the product has different (independent) features, the FMI also makes it possible, depending on requirements, to obtain partial information for each feature.

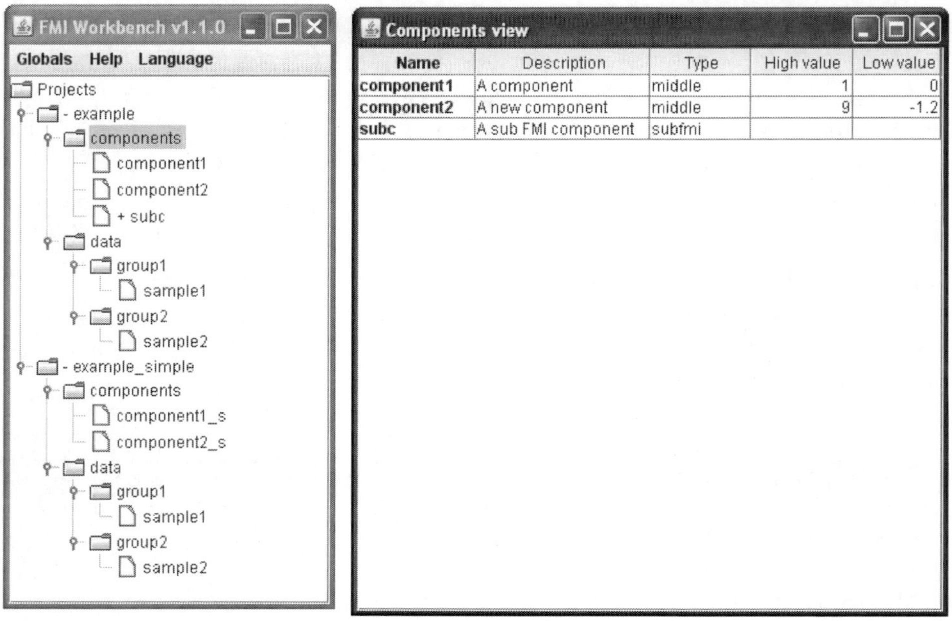

Figure 14.

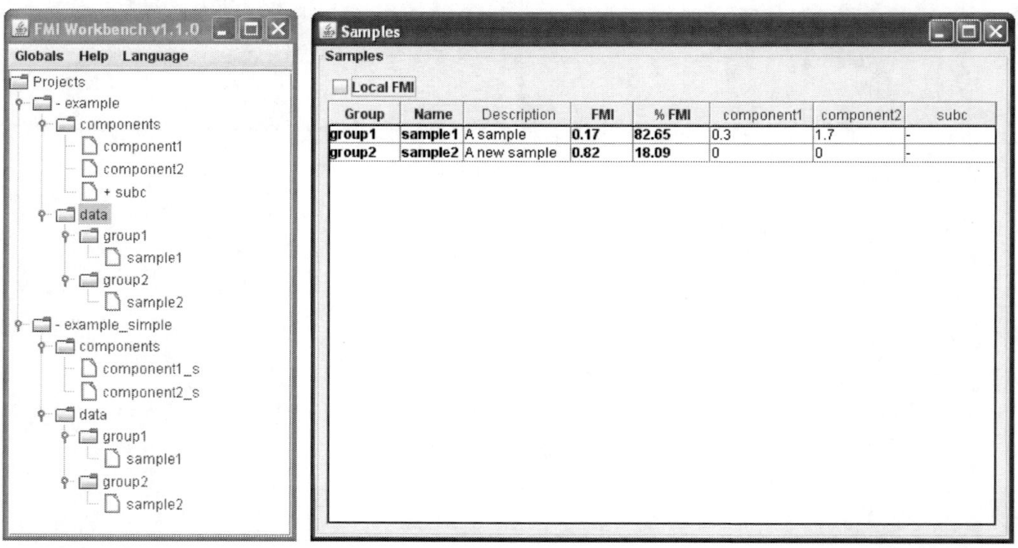

Figure 15.

Thanks to the parameters normalization and the introduction of the indexes, in the FMI calculation we can arbitrarily link different parameters, because we compare pure numbers of the same order of magnitude, instead of different experimental values in different unit measures. Thus the index can predict and classify beneficial (biological, nutritional etc.) effects of foods in terms of content of structurally different compounds, because, paraphrasing Purcell [31], quality, and consequently the FMI, cannot be reduced to a simple sum of experimental values, but is given by the relationships, the synergies, the interactions among different compounds and components, which can be adequately captured and represented by the FMI.

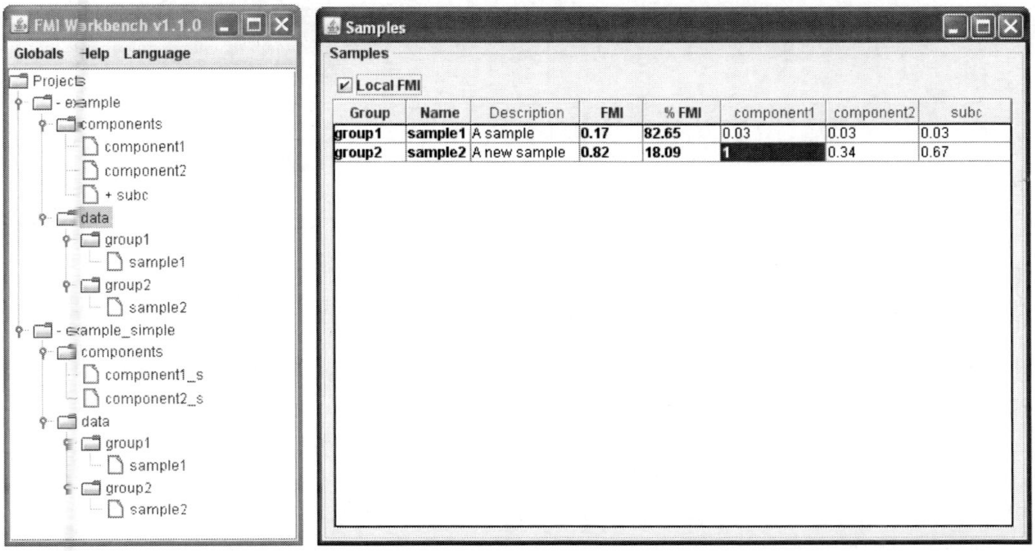

Figure 16.

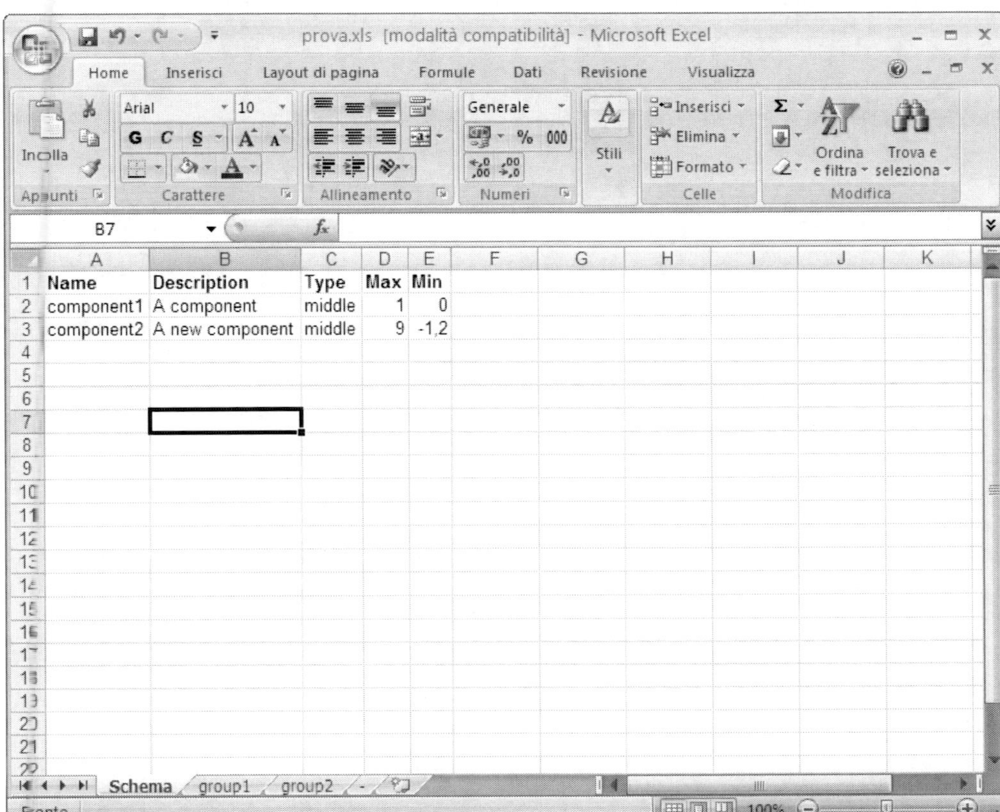

Figure 17.

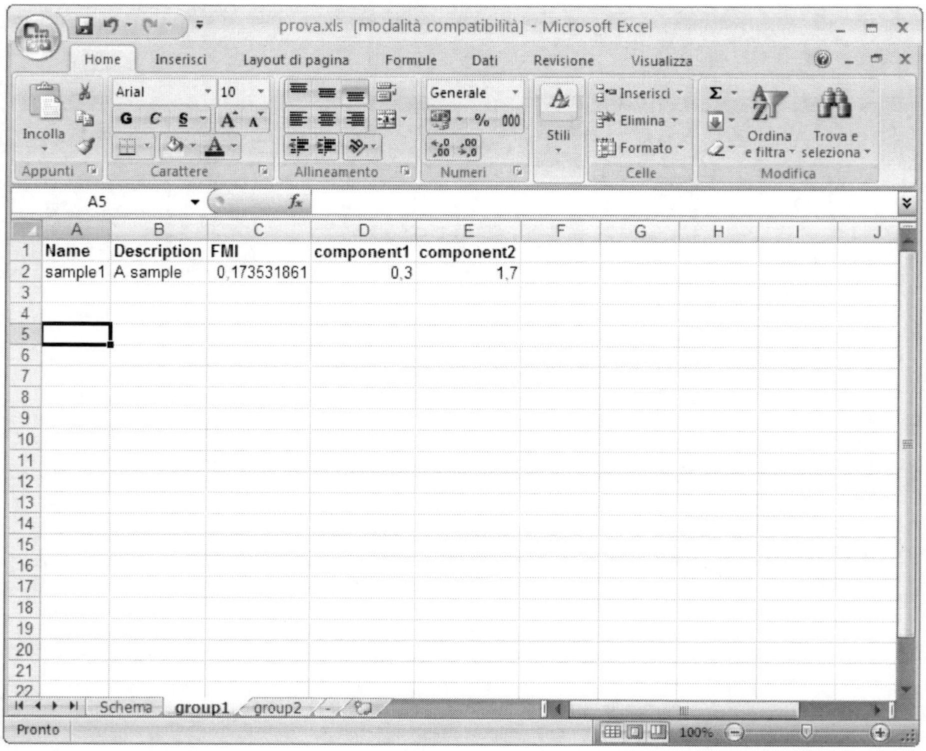

Figure 18.

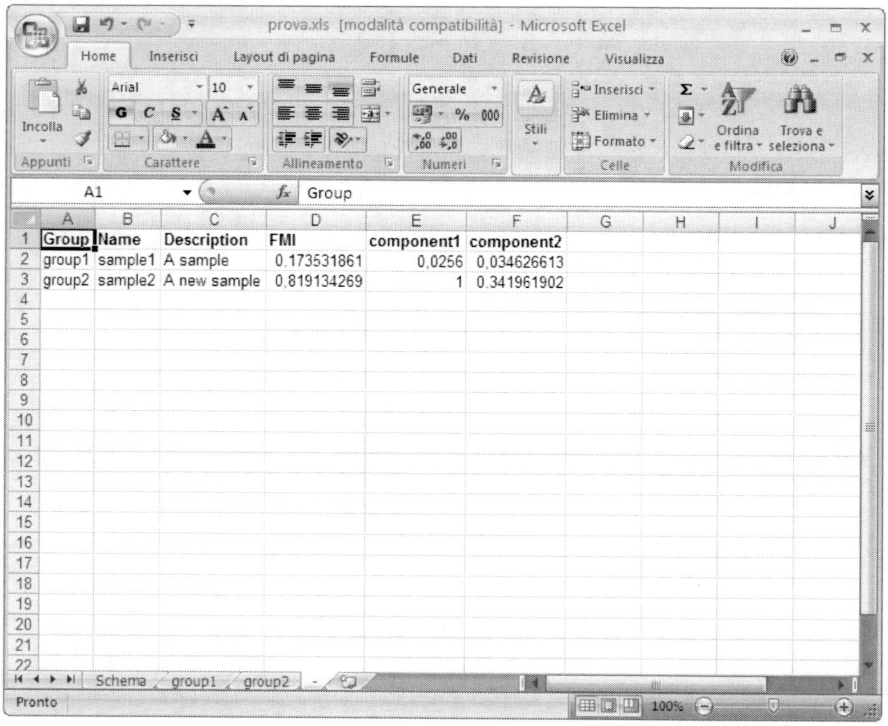

Figure 19.

In conclusion the FMI is a tool which can help a mathematical approach to quality, based on the observation and quantification, by a synthetic value, of different (biological, chemical, clinical etc.) aspects of products and processes and could be applied also in other fields, where the "measure" of some features can be given only in terms of discrete variables, roughly corresponding to goodness categories (bad, good, mediocre etc.)

Finally, it is important to note that the computational tool of the FMI index is not alternative to classical statistical methods (multivariate analysis, cluster analysis, etc.), but in some sense it could complement such analyses.

REFERENCES

[1] Pomeranz, Y. (1991). *Functional properties of Food Components.* 2nd Edition. Academic Press, San Diego, U.S.A.

[2] Brody, T. (1994). *Nutritional Biochemistry*. Academic Press, San Diego, U.S.A.

[3] FAO/WHO (1989). *Protein quality evaluation*, Report of the joint FAO/WHO Expert Consultation, 51, Bethesda, Md., U.S.A.

[4] Potter, N.N.& Hotchkiss, J.H. (1995). *Food Science* 5th Edition, Chapman and Hall, New York, U.S.A..

[5] Nawar, W.W. (1996). Lipids. In *Food Chemistry,* Fennema O.R. Ed., Marcel Dekker Inc., New York-Basel, pp. 225-319.

[6] Gregory III, J.F. (1996). Vitamins. In *Food Chemistry,* Fennema O.R. Ed., Marcel Dekker Inc., New York-Basel, 531-616.

[7] Miller, D.D. (1996). Minerals. In *Food Chemistry,* Fennema O.R. Ed., Marcel Dekker Inc., New York-Basel, 617-649.

[8] Woyengo, T.A.& Nyachoti, C.M. (2013). Review: Anti-nutritional effects of phytic acid in diets for pigs and poultry - current knowledge and directions for future research. *Can. J. Anim. Sci.*, 93, 9-21.

[9] Krupa, U. (2008). Main nutritional and antinutritional compounds of bean seeds - a review. *Pol. J. of Nutr. Sci.*, 58, 149-155.

[10] Doria, E., Campion, B., Sparvoli, F., Tava, A.& Nielsen, E.(2012). Anti-nutrient components and metabolites with health implications in seeds of 10 common bean (Phaseolus vulgaris L. and Phaseolus lunatus L.) landraces cultivated in southern Italy. *J. Food Comp. Anal.*, *26*, 72–80.

[11] Shahidi, F. (1997). Beneficial Health Effects and Drawbacks of Antinutrients and Phytochemicals in Foods. An Overview. In *Antinutrients and Phytochemicals in Food*, ACS Symposium Series, *American Chemical Society*, Washington, U.S.A., 1-9.

[12] Friedman, M.& Levin, C.E. (2008). Review of methods for the reduction of dietary content and toxicity of acrylamide. *J. Agr. Food Chem.*, *56*, 6113-6140.

[13] Schroder, M.J.A. (2003). *Food Quality and consumer value.* Springer-Verlag, Berlin, Germany, 53-88.

[14] Green Paper, E. U. (2008). European Commission, Bruxelles.

[15] Heiberg, J. L. (1881). Archimedis opera omnia III, Prolegomena XCV, Teubner, Leipzig, Germany.

[16] Finotti, E., Bersani, A.M.& Bersani, E. (2007). Total quality indexes for extra virgin olive oils. *J. Food Qual.*, *30*, 911-931.

[17] Finotti, E., Bersani, E., Bersani, A.M., Vivanti, V.& Toti, E. (2010). Variation of nutritional parameters in the olive oil during the ripening phase. La Rivista di Scienza dell'Alimentazione. Anno 40, numero 11, 33-41.

[18] Jordan, D.W.& Smith, P. (1994). *Mathematical Techniques*, Oxford University Press, Oxford, U.K., 7.

[19] Horn, R.A.& Johnson, C.R. (1990). *Matrix Analysis*, Cambridge University Press, West Nyack, U.S.A., 290–320.

[20] Finotti, E., Bersani, E., Vivanti, V.& Friedman, M. (2008). Application of a functional mathematical index to the evaluation of the nutritional quality of potatoes. In *Potato IV, FOOD 3 (Special Issue 2)*, Yee N., Bussell W. Eds., Global Scientific Books, Isleworth, U.K., 30-36.

[21] Deza, E.& Deza, M.M.(2009). *Encyclopedia of Distances.* Springer, New York, U.S.A., 94.

[22] Finotti, E., Bersani, E., Del Prete, E.& Friedman, M. (2012). A Functional mathematical index for predicting effects of food processing on eight sweet potato (Ipomoea batatas) cultivars. *J. Food Comp. Anal.*, *27*, 81-86.

[23] Finotti, E., Bersani, E., Vivanti, V.& Friedman, M. (2011). Application of a functional mathematical quality index to asparagine, free sugar and phenolic acid content of 20 commercial potato varieties. *J. Food Qual.*, *34*, 74-79.

[24] Finotti, E., Bersani, E., Vivanti, V.&Friedman, M. (2010). *Application of a Functional Mathematical Index (FMI) to Composition of Potato Glycoalkaloids that Predicts Food Quality and Safety, La Rivista di Scienza dell'Alimentazione,* anno 39, numero 3, 27-38.

[25] Finotti,E., Bersani, E. & Friedman, M. (2011). Application of a Functional Mathematical Index for Antibacterial and Anticarcinogenic Effects of Tea Catechins, *J. Agric. Food Chem., 59,* 864–869.

[26] Finotti,E., Bersani, E., Del Prete, E. & Friedman, M. (2013). Application of a functional mathematical index (FMI) for predicting effects of the composition of jujube fruit on nutritional quality and health, submitted to *J. Food Comp. Anal.*

[27] Bersani, A.M., Finotti, E., Bersani, E.& Del Prete, E. (2011). Uso di un indice matematico funzionale (FMI) per la misura della qualità dell'acqua potabile, ai fini di un corretto rapporto costi/qualità. Accademia Nazionale dei Lincei, Atti dei Convegni Lincei 266. XI Giornata Mondiale Dell'Acqua, Convegno "Acqua ed Energia". Roma 22 Marzo, 163-170.

[28] Finotti, E., Bersani, A.M., Bersani, E.& Friedman, M. (2011). La formazione dell'acrilamide nelle patate fritte e scelta delle cultivar più idonee mediante indice matematico funzionale", La Rivista di Scienza dell'Alimentazione (2011), supplemento al n. 1/2011. Atti del Convegno "Il processo di frittura: ricerca e innovazione", 90-98.

[29] Gamma, E., Helm, R., Johnson, R. & Vlissides, J. (1997). Design Patterns: Elements of Reusable Object-Oriented Software. Addison-Wesley, Boston, U.S.A..

[30] Fowler, M. (2003). UML Distilled: A Brief Guide to the Standard Object Modeling Language (3rd Edition), Addison-Wesley, Boston, U.S.A.

[31] Purcell, E.M. (1971). La parte e il tutto in fisica. In *Qualità e quantità e altre categorie della Scienza,* Lerner D. Ed., Boringhieri, Torino, Italy, 164-177.

In: Processed Foods
Editors: Chloe M. Gagne and Daniel B. Jones

ISBN: 978-1-62808-688-1

Chapter 6

NUTRITIONAL AND BIOLOGICAL POTENTIAL OF FUNCTIONAL FOODS CONTAINING THE HARD-TO-COOK BEAN (*PHASEOLUS VULGARIS* L.) PROTEIN HYDROLYSATE

Maira Rubi Segura Campos, Karem García-Rodríguez, Luis Chel Guerrero and David Betancur Ancona*

Facultad de Ingeniería Química, Universidad Autónoma de Yucatán, Mérida, Yucatán, México

ABSTRACT

Advances in understanding of the relationship between nutrition and health contributed to development of the concept of functional foods, a practical and new approach to achieving optimal health through promoting well-being and possibly reducing disease risk. Functional foods are found in virtually all food categories. Pasta is considered a functional food for its low fat and sodium contents. Wheat semolina is highly suitable for making pasta because its proteins are ideal for dough development and for preventing pasta disintegration during cooking. However, wheat pasta is nutritionally unbalanced due to its low protein content (<15%) and relatively deficient levels of the essential amino acid lysine. Wheat pasta is an effective vehicle for food supplementation with minerals, proteins and other health-promoting components. The present chapter objective was to evaluate the nutritional and biological potential of semolina pasta added with hard-to-cook bean (*Phaseolus vulgaris* L.) protein hydrolysate. Hard-to-cook bean protein hydrolysate obtained with the Alcalase®-Flavourzyme® sequential enzymatic system was added to durum wheat semolina pasta at two concentrations (5 and 10%). The pastas exhibiting angiotensin I-converting enzyme (ACE) inhibitory activity (10.88-17%) and antioxidant activity (TEAC = 26.27-31.42 mM/mg protein). Evaluated nutritional parameters and sensory scores indicated the best replacement level to be 10% hydrolysate. Addition of a hard-to-cook bean hydrolysate to semolina pasta produced by extrusion improved product nutritional value without significantly affecting sensory

* Telephone: 52 999 946-09-56, *Email: bancona@uady.mx.

parameters. Semolina pasta containing 10% protein hydrolysate from hard-to-cook *P. vulgaris* can be considered a functional food with both biological activity and a good nutritional profile.

Keywords: Hard-to-cook bean, protein hydrolysates, biological activities, nutritional evaluation

INTRODUCTION

The human body is constantly subjected to physiological imbalances and exposure to extrinsic toxic substances that perturb normal functions leading to various health conditions. These aberrations can be controlled by physiological homeostasis, or through the use of health-promoting agents especially in acute and chronic conditions. It is generally established that the nutritive and non-nutritive constituents of food can be used to modify the risk of developing or aggravating human disease conditions. In this regard, functional foods and nutraceuticals have emerged as adjuvant or alternative to chemotherapy especially in prevention and management of human diseases, and for maintaining optimum health state (Udenigwe and Aluko, 2012). Food-derived bioactive peptides represent a source of health enhancing components that may be incorporated in functional foods and in pharmaceutical preparations. These peptides are potential modulators of various regulatory processes in the body (Meisel *et al.*, 2001). Legumes proteins are precursors of many different biologically active peptides, e.g. peptides having antioxidant, hypotensive, antithrombotic or antimicrobial activities. The intrinsic bioactivities of the peptides encrypted in major legumes proteins are latent until they are released and activated by enzymic hydrolysis, e.g. during gastrointestinal digestion and/or food processing (Meisel *et al.*, 2001). The physiological effect of biopeptides depends on their ability to reach the target site. Biofunctional peptides have the potential to be applied in the formulation of health enhancing-nutraceuticals with well-defined pharmaceutical effect. Peptide products which have ACE inhibitory effect are currently on the market and some of them are at the stage of being tested due to the strict requirements in order to demonstrate the efficacy of these bioactive peptides prior to their widespread utilization as physiologically beneficial functional foods/food ingredients (Iwaniak and Minkiewicz, 2007). According to Korhonen and Pihlanto (2007) there is a need to develop methods to optimise the activity of bioactive peptides in food systems and to enable their optimum utilization in the body. Many food and pharmaceutical companies consider the exploitation of bioactive peptides in human nutrition and health promotion. Although the consumers knowledge about the impact of foods on the body rise, the molecular studies as well special processing protocols are required before the application of biofunctional peptides as the food components (Iwaniak and Minkiewicz, 2007). The European Comission Concerted Action on Functional Food Science in Europe (FUFOSE) proposed a working definition of functional food: a food that beneficially affects one or more target functions in the body beyond adequate nutritional effects in a way that is relevant to either an improved state of health and well-being and/or reduction of risk of disease. It is consumed as part of a normal food pattern. It is not a pill, a capsule or any form of dietary supplement. While there is increased awareness of the links between diet and disease such as certain fats and CVD, calcium and osteoporosis, fibre and gastrointestinal (GI) health, it is important, for a functional food, to identify the specific food constituents that could promote health and well-being as well as the exact conditions where they can have this beneficial effect (European commissión, 2010). The Japanese scientific academic community defined 'functional food' early in the 1980s. That is, functional foods are those that have three functions. The primary function is nutrition. The secondary function is a sensory function or sensory satisfaction. The third is the tertiary function, which is physiological. The Japanese Ministry of Health, Labour, and Welfare (MHLW) set up 'Foods for Specified Health Use' (FOSHU) in 1991 as a regulatory system to approve the statements made on food labels concerning the effect of the food on the human body. Food products applying for approval by FOSHU are scientifically evaluated in terms of their effectiveness and safety by the Council of Pharmaceutical Affairs and Food Hygiene under the MHLW. The regulatory range of FOSHU was broadened in 2001 to accept the forms of capsules and tablets in addition to those of conventional foods. FOSHU increased the total to about 330 items in January 2003 (Shimizu, 2003).

Most of the FOSHU products with bloods pressure lowering effects in humans contain ACE inhibitory peptides as the active component. These active peptides are relatively short, rich en Pro at the C-terminus and contain branched amino acids. Ameal S® was initially prepared from *Lactobacillus helveticus* fermented milk; however the active peptides, Val-Pro-Pro and Ile-Pro-Pro are now routinely prepared by enzymatic hydrolysis of casein by an *Aspergillus oryzae* proteases. The casein hydrolysate was used recently in a new vegetable juice and approved as a FOSHU product. In the United States, the casein hydrolyzate containing Val-Pro-Pro and Ile-Pro-Pro has been used in a supplement product, Ameal bp®. Valio, a Finnish company, launched a fermented milk product containing Val-Pro-Pro and Ile-Pro-Pro in Finland in 2002. Blood pressure-lowering effects in Finnish people drinking the fermented milk product have been proven (Seppo *et al.*, 2002). Since then, similar products with different Brand names have been launched in various European countries, MS LH (*L. helveticus*) in Iceland (2004), Kaiku Vita in Spain (2005) and Emmi Evolus in Portugal and Switzerland (2005). Other peptides derived from food proteins and showing significant health benefits in Japanese people were approved as FOSHU products in Japan. Val-Tyr with ACE inhibitory activity was prepared from sardine proteins by the proteolytic action of a *Bacillus* alkaline protease (Matsui *et al.,* 1993). Leu-Lys-Pro-Asn-Met is prepared by hydrolysis of bonito with thermolisin (Yokoyama *et al.,* 1992). Leu-Val-Tyr was prepared from sesame proteins by food-grade enzymes. In the manufacture of biogenic peptides, commercially available food-grade enzymes to hydrolyze food proteins are adventageous because they are cheap and safe. Pasta is considered a functional food for its low fat and sodium contents. Wheat semolina is highly suitable for making pasta because its proteins are ideal for dough development and for preventing pasta disintegration during cooking. However, wheat pasta is nutritionally unbalanced due to its low protein content (<15%) and relatively deficient levels of the essential amino acid lysine (Shogren *et al.*, 2006). Wheat pasta is an effective vehicle for food supplementation with minerals, proteins and other health-promoting components (Borneo and Aguirre, 2008). Inadequate post-harvest handling and storage under high temperature (>25 °C) and relative humidity (>65%) conditions produce the hard-to-cook (HTC) defect in beans. Alkaline extraction is a technological alternative for protein isolation from HTC beans. The isolated protein has low trypsin inhibitor activity and meets FAO recommendations for essential amino acids content in diets for adults, meaning it is potentially useful as a supplementary vegetable protein source or for bioactive peptides for using in food manufacturing (Morales-De-León et al., 2007).

The present study objective was to incorporate hard-to-cook bean (*Phaseolus vulgaris* L.) protein hydrolysate into semolina pasta and evaluate the biological and nutritional potential of the resulting products.

MATERIALS AND METHODS

Seeds and Chemicals

Common beans (*Phaseolus vulgaris* L.) (10 kg) were stored for 4 months under uncontrolled temperature and relative humidity conditions. All chemicals were reagent grade and purchased from Sigma Chemical Co. (St. Louis, MO, USA).

Enzymatic Hydrolysis

A single extraction was done with 10 Kg HTC bean flour using a wet fractionation method to obtain the protein concentrate (Ruiz-Ruiz *et al.*, 2012). Using the sequential enzymatic system Alcalase®/Flavourzyme® (Novo Nordisk, Bagsvaerd, Denmark), hydrolysis was done under controlled conditions in a 1000 mL reaction vessel equipped with a stirrer, thermometer and pH electrode (Pedroche *et al.*, 2002). Protein concentrate was suspended in distilled water to produce a 10% (w/v) protein solution that was equilibrated at optimum temperature and pH for each protease before adding the respective enzyme. Protease was then added to the solution at a ratio of 0.3 AU/g for Alcalase® and 50 LAPU/g for Flavourzyme®. Hydrolysis conditions were 45 min at 50 °C for each enzyme, using pH 8.0 for Alcalase® and pH 7.0 for Flavourzyme®. The pH was kept constant by adding 1.0 M NaOH during hydrolysis. The reaction was stopped by heating to 80 °C for 20 min. Degree of hydrolysis (DH) was calculated by determining free amino groups with o-phthaldialdehyde (Nielsen *et al.*, 2001): $DH = h/h_{tot} * 100$; where h_{tot} is the total number of peptide bonds per protein equivalent, and h is the number of hydrolyzed bonds. The h_{tot} factor is dependent on raw material amino acid composition. For protein isolate from HTC beans h_{tot} = 7.66 mmol/g protein.

Extrusion Process

Durum wheat semolina samples containing four different levels of HTC *P. vulgaris* hydrolysate (0, 5, 10 and 15%) were extruded using a Brabender 10 DN monoscrew extruder with the following specifications: screw without compression ratio; three 1.5 mm diameter x 20 mm long (1.5 x 20) die orifices and a pressure transducer (Dynisco, Model PT420) with a measuring range of 0 to 350 kg/cm^2. Extrusion conditions were 30 °C, 34% moisture content and 100 rpm screw speed. After extrusion, the products were placed on trays to cool for 10 min. Product moisture content was conditioned for 22 h at 39 °C and 75% relative humidity, and products were stored in polyethylene bags until analysis.

Biological Activities

Protein hydrolysate and pastas containing the hydrolysate were analyzed to identify and quantify ACE inhibitory and antioxidant activities.

ACE Inhibitory Activity

Angiotensin I-converting enzyme inhibitory activity in the hydrolysates was analyzed following Hayakari and Kondo (1978). Hippuryl-L-histidyl-L-leucine (HHL) is hydrolyzed by ACE to yield hippuric acid and histidyl-leucine. This method relies on the colorimetric reaction of hippuric acid with 2,4,6-trichloro-s-triazine (TT) in a 0.5 ml incubation mixture containing 40 μmol potassium phosphate buffer (pH 8.3), 300 μmol sodium chloride, 40 μmol 3% HHL in potassium phosphate buffer (pH 8.3) and 100 mU/ml ACE. The mixture was incubated at 37 ºC for 45 min and the reaction terminated by adding TT (3% v/v) in dioxane and 3 ml 0.2 M potassium phosphate buffer (pH 8.3). After centrifuging the reaction mixture at 10,000 *x g* for 10 min, enzymatic activity in the supernatant was measured at 382 nm absorbance. All runs were done in triplicate. ACE inhibitory activity was quantified by a regression analysis of ACE inhibitory activity (%) versus peptide concentration and defined as an IC_{50} value, that is, the peptide concentration (μg protein/mL) required to produce 50% ACE inhibition under the described conditions.

$ABTS^{\bullet+}$ Decolorization Assay

The $ABTS^{\bullet+}$ radical cation was produced by reacting 2,2'-azino-bis(3-ethylbenzthiazoline-6-sulphonic acid (ABTS) with potassium persulfate (Pukalskas *et al.*, 2002). To prepare the stock solution, ABTS was dissolved at a 2 mM concentration in 50 mL phosphate-buffered saline (PBS) prepared from 4.0908 g NaCl, 0.1347 g KH_2PO_4, 0.7098 g Na_2HPO_4, and 0.0749 g KCl dissolved in 500 mL ultrapure water. A 70 mM $K_2S_4O_8$ solution in ultrapure water was prepared. The ABTS radical cation was produced by reacting 10 mL ABTS stock solution with 40 μL $K_2S_4O_8$ solution and allowing the mixture to stand in darkness at room temperature for 16-17 h before use. Antioxidant compound content in the hydrolysates was analyzed by diluting the $ABTS^{\bullet+}$ solution with PBS to an absorbance of 0.800 ± 0.030 AU at 734 nm. After adding 990 μL diluted $ABTS^{\bullet+}$ solution (A 734 nm= 0.800 ± 0.030) to 10 μL 6-hydroxy-2,5,7,8-tetramethylchroman-2-carboxylic acid (TROLOX) standard (final concentration 0.5-3.5 mM) in PBS, absorbance was read at room temperature exactly 6 min after initial mixing. The percentage decrease in absorbance at 734 nm was calculated and plotted as a function of the antioxidant concentration, or of Trolox for the standard reference data. To calculate the Trolox equivalent antioxidant coefficient (TEAC), the slope of the absorbance inhibition percentage vs. antioxidant concentration plot was divided by the slope of the Trolox plot. This produces the TEAC at a specific point in time.

Nutritional Evaluation

Nutritional quality analyses were done of the pasta treatment with the best physical characteristics and highest biological activity.

Proximate Composition

Proximate composition of the HTC bean flour, protein concentrate and pastas containing HTC bean hydrolysates was determined using official AOAC procedures (AOAC, 1997): nitrogen (method 954.01); fat (920.39); ash (923.03); fiber (962.09) and moisture (925.09). Protein content was calculated as nitrogen x 6.25, and carbohydrate content estimated as nitrogen-free extract (NFE).

In Vitro Protein Digestibility

Following Rudloff and Lönnerdal (Rudloff and Lönnerdal, 1992), 10 ml pasta suspension were prepared with an enzyme/substrate ratio of 1:15 (w/w) based on protein content. The pH was adjusted to 4.5 with 1.0 M HCl and pepsin added at a 10 mg/sample concentration. The suspension was incubated in darkness in a shaking water bath at 37 °C for 30 min. After incubation, pH was gradually raised to 7 using 0.5 M $NaHCO_3$. A volume of 2.5 mL pancreatin solution (0.4g/100mL in 0.1 M $NaHCO_3$) was added and incubated for 1 h at 37 °C. Enzymes were inactivated by placing samples in boiling water for 4 min. The suspension was centrifuged at 272,000 *g* for 45 min at 4 °C to separate the lipid, soluble and insoluble fractions. Protein digestibility is defined as the increase in non-protein nitrogen (NPN) after enzymatic digestion. NPN was measured in the soluble portion and the undigested sample, using trichloroacetic acid (24%) at a 1:1 ratio. The solution was then centrifuged at 6000 *g* and the supernatant removed. Nitrogen content was measured by the Kjeldahl method (AOAC, 1997) and calculated as follows:

NPN = NPN digested sample - NPN undigested sample

Digested protein = (NPN) (6.25)

Digestibility (%) = (Protein digested x 100) / Total protein

Available Lysine

Available lysine content was determined with the method of Hurrel *et al.* (1979) Briefly, 0.5 g sample was weighed into two tubes (A, B), 1 mL 2-propanol added to each tube and these agitated vigorously for 5 min. After agitation, 4 mL sodium acetate (5%) was added to tube A, and 4 mL sodium acetate plus 0.3 mL propionic anhydride added to tube B.

The tubes were shaken vigorously for 15 min, 40 mL coloring solution added to each, and then shaken again for 60 min. Excess coloring concentrations were determined at 475 nm using a spectrophotometer (Thermospectronic Genesis 10uv, Madison, WI, USA). Lysine levels were determined by the B - A difference. In both tubes, millimoles bound coloring = (40 mL/1000 mL) (100/weight of sample) (0.146), where 0.146 = lysine conversion factor.

Amino Acid Content

Amino acid profiles were generated according to Alaiz *et al.* (1992) using precolumn derivatization with diethyl ethoxymethylenemalonate and reverse-phase high-performance liquid chromatography (RP-HPLC) with spectrophotometric detection at 280 nm. The HPLC system (Waters) consisted of a model 600E multi-solvent delivery system, a Wisp Model 712 automatic injector and a model 484 UVevis detector. Amino acid derivative resolution was constantly determined using a binary gradient system. Solvents were (A) 25 mmol/L sodium acetate containing 0.02% sodium azide (pH 6.0) and (B) acetonitrile. Solvents were injected into the column at a 0.9 mL/min flow rate at the following intervals: time 0.0-3.0 min, linear gradient from A-B (91:9) to A-B (86:14); 3.0-13.0 min, elution with A-B (86:14); 13.0-30.0 min, linear gradient from A-B (86:14) to A-B (69:31); and 30.0-35.0 min, elution with A-B (69:31).

Protein Efficiency Ratio (cPER)

This parameter was calculated following the applicable AOAC (1997) method, employing the *in vitro* digestibility value and the amount (g) of amino acid (AA)/100 g protein of Lys, Met + Cys, Thr, Ile, Leu, Val, Phe + Tyr and Trp. This assumes that the Cys and Tyr values in the Met + Cys and Phe + Tyr combinations cannot surpass 50% of the total of their respective combinations.

Total Starch Content

This was quantified by adapting the methodology of Tovar *et al.* (1990) using 4.0 M KOH for starch solubilization. Hydrolysis was done using thermostable α-amylase (Sigma TO-3306) and amyloglucosidase (BioChemika 10115 70 solid units/mg). Reactivated glucose oxidase/peroxidase (GODePAP) (DiaSys, RE. 10250021) was used for colorimetric determination of glucose, and the concentration measured at 500 nm in a spectrophotometer (Thermospectronic Genesis 10UV, Madison, WI, USA). Calculation of sample glucose concentration was done using the equation: glucose [mg/dL] = (SA - BA/StA - BA) (StC), followed by %Starch = (mg glucose x 2 x 0.9/sample weight) (100); where, SA = Sample absorbance; BA = Blank absorbance; StA = Standard absorbance; StC = Standard concentration = 100 mg/dL; 2 = dilution factor; and 0.95 = glucose to glean transformation factor.

Available Starch Content

Adapting the methodology of Holm *et al.* (1986) this was quantified in the same way as total starch, but 4.0 M KOH was not added.

Resistant Starch

This parameter was calculated by the difference between Total Starch (TS) and Available Starch (AS): Resistant starch (RS) = TS – AS (Tovar *et al.*, 1990).

In Vitro Starch Digestibility

In vitro hydrolysis was determined using the method of Holm *et al.* (1986) which is based on the reducing power of the maltose released by the action of pancreatic amylase through 3,5-dinitrosalicylic acid (DNS). A standard maltose curve was generated and the calculations done as follows: %Hydrolysis in time 0: %Hydrolysis = [(mg maltose to 0 min - mg maltose 0 m) x 0.95 x 100]/1.82, where 0 m is a blank mixture of 0.3 mL H_2O, 0.5 mL standard, 1 mL DNS and 0.2 mL of sample; %Hydrolysis 5-60 min: %Hydrolysis = [mg maltose - (mg maltose to 0 min - 1.0)] x 0.95 x 100]/1.78, where, 1.0 is subtracted because 1.0 mg maltose was added to the 0 min and 0 m samples to ensure their detection in the spectrophotometer, 0.95 = glucose to glean transformation factor, and 5-60 = minutes hydrolysis times.

Total Dietary Fiber

This parameter was determined with the gravimetric enzymatic method proposed by Prosky *et al.* (1998) Briefly, 1 g sample was weighed into each of four flasks, and 50 mL 0.05 N phosphate buffer at pH 6 added to each. The flasks were then placed in a Galena bath at 100 °C, 0.1 mL thermostable α-amylase enzyme (Sigma A-3306) added to each and then they were agitated at 60 rpm for 15 min. After cooling, pH was adjusted to 7.5. The flasks were returned to the bath at 60 °C, 0.1 mL protease (Sigma P-3910) added to each and then they were agitated at 60 rpm for 30 min. After cooling, pH was adjusted to 4.0. The flasks were again placed in the bath at 60 °C, 0.3 mL amyloglucosidase (Sigma A-9913) added and then they were agitated for 30 min. Finally, 95 g/kg ethanol, preheated to 60 °C, was added at a 1:4 (v/v) ratio. In a vacuum, flask content was filtered into crucibles containing celite (Sigma C-8656). The residue remaining in the flask was washed three times with 20 mL 780 g/kg ethanol, twice with 10 mL 950 g/kg ethanol and twice with 10 mL acetone. Crucible content was dried at 105 °C. Residue protein content (N x 6.25) was calculated for two crucibles, and the residue in the two remaining crucibles was burned at 550 °C for 4 h.

TDF (%) = [Residue weight (g) - protein (g) - ash (g) - Blank]/ Sample weight (g) x 1000

Insoluble Dietary Fiber

This analysis is similar to that for total dietary fiber (TDF), except that addition of 95% alcohol at 1:4 is omitted (Prosky *et al.*, 1998).

Soluble Dietary Fiber

Soluble dietary fiber (SDF) was calculated by the difference between TDF and insoluble dietary fiber: SDF = TDF – IDF (Prosky *et al.*, 1998).

Sensory Evaluation

Sensory evaluation was done using cooked pasta samples and a panel of eighty untrained judges. Sensory properties were evaluated on a 7-point scale: 7 = like very much; 6 = like; 5 = like a little; 4 = neither like nor dislike; 3 = dislike a little; 2 = dislike; 1 = dislike very much. Sample preparation was done by cooking 25 g pasta in 950 mL water containing 2 g NaCl and 15 g butter for the optimum time (8 min). Once cooked, it was drained and served to the judges (Bergman *et al.*, 1994).

Statistical Analysis

All results were analyzed using descriptive statistics with a central tendency and dispersion measures. One-way ANOVAs were run to evaluate physical characteristics, nutritional contribution, sensory acceptance, and biological (antihypertensive and antioxidant) capacities. A LSD multiple range test was applied to calculate differences between treatments. All analyses were done according to Montgomery (2004) and processed with the Statgraphics Plus version 5.1 software.

RESULTS AND DISCUSSION

Degree of Hydrolysis and Biological Activity of HTC Bean Protein Hydrolysate

Crude protein content in the HTC bean flour was 21.7%. Using alkaline extraction and isoelectric precipitation, HTC bean protein isolate crude protein content was 73.03%. The Alcalase®-Flavourzyme® sequential enzymatic system produced a high DH (43.01%). The non-hydrolyzed HTC bean protein concentrate showed no biological activity, in contrast, enzymatic hydrolysis produced hydrolysates with bioactivity. The Alcalase®-Flavourzyme® hydrolysate (AFH) had an IC_{50} of 5460.12 μg protein/mL, much higher than reported for enzymatic hydrolysates from different protein sources (IC_{50} = 200 to 2467 μg protein/mL) with antihypertensive activity in spontaneously hypertensive rats (Hong *et al.*, 2005). This

Total, Available and Resistant Starch Contents

Total starch was lower in the 10% AFH treatment (71.1%) than in the control (82.9%), a 14% reduction in starch content proportional to the hydrolysate inclusion level. The available starch/resistant starch ratio (3.5) was the same in both pastas. Consumption of the 10% AFH pasta could reduce caloric intake from starch and increase the nutritional contribution of protein, making this product a functional food.

In Vitro Starch Digestibility

In vitro starch digestibility was evaluated at five moments over a 60 min period (Figure 4). The digestibility kinetic of the control increased more slowly ($p<0.05$) than the 10% AFH treatment, but once the 10% AFH treatment reached 67.7% (5 min) it increased at a lower rate, providing a more controlled increase in glycaemia. One possible cause is the low extrusion speed (100 rpm) used in the present study since low speeds allow starch solubilization and gelatinization, thus improving digestibility (Dahlin and Lorenz, 1993). Indeed, starch digestion rate is useful for predicting postprandial glycemic response, with rapidly digested starches more quickly increasing glycaemia (Osorio *et al.*, 2004).

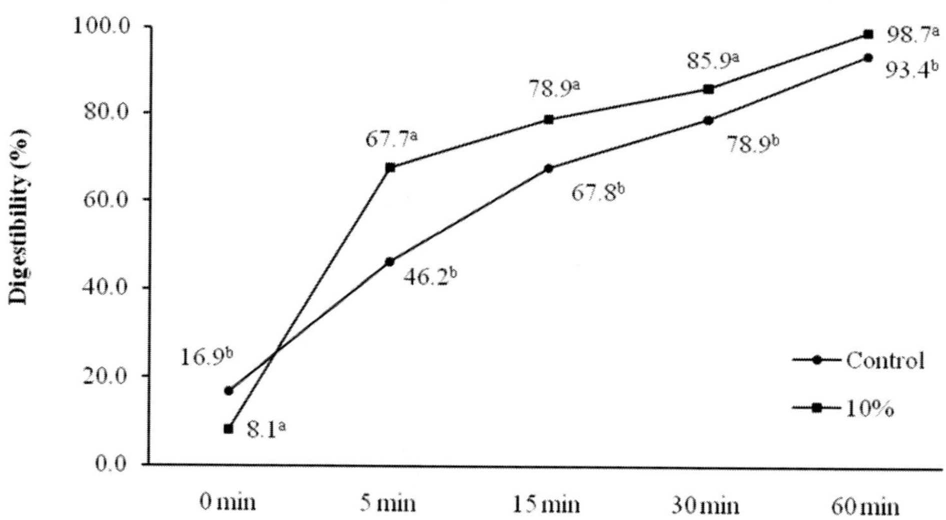

Figure 4. *In vitro* starch digestibility (%) of wheat semolina pasta (control) and a treatment containing 10% hard-to-cook bean (*Phaseolus vulgaris* L.) protein hydrolysate. Different letters indicate statistical difference ($p<0.05$).

Total, Insoluble and Soluble Dietary Fiber

Dietary fiber values varied minimally between the control and the 10% AFH treatment (Table 2). Dietary fiber has been widely studied for its health benefits. It is considered a preventive factor for cancer, serves as a substrate for colonic bacteria, promotes intestinal food transit, and decreases bile acid reabsorption thereby altering micelle formation and contributing to lowering blood cholesterol levels (Escudero Alvarez and González Sánchez,

2006). The 10% AFH treatment retained an adequate insoluble/soluble dietary fiber ratio, so its consumption would provide the benefits of fiber intake.

Sensory Evaluation

The 10% AFH treatment was well accepted by the judges (73.4% of scores from Like very much, Like and Like a little) (Figure 5). Above a 10% legume protein substitution level, consumers due to a beany taste may reject products and undesirable changes in mouth feel (Bahnassay and Khan, 1986).

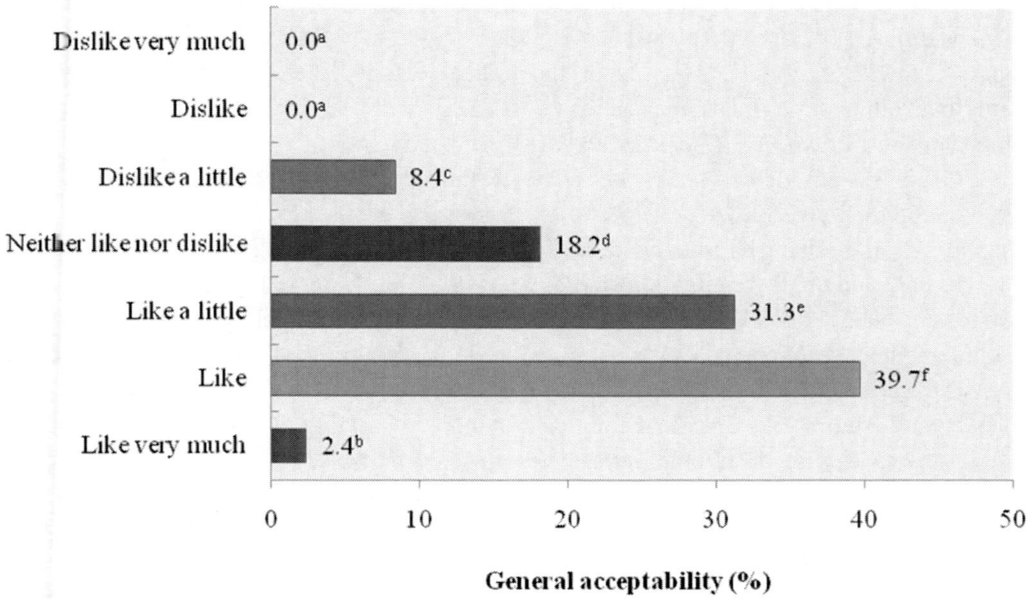

Figure 5. General acceptability (%) of pasta containing 10% hard-to-cook bean (*Phaseolus vulgaris* L.) protein hydrolysate. Different letters indicate statistical difference ($p<0.05$).

Conclusion

Hard-to-cook bean proteins are no longer considered merely nutritional components because they possess encrypted peptides with ACE-inhibitory and antioxidant activities. Semolina pasta containing 10% protein hydrolysate from hard-to-cook *P. vulgaris* can be considered a functional food with both biological activity and a good nutritional profile. These findings open an industrial perspective that will permit the unrestricted use of hard-to-cook bean protein hydrolysates in healthy promoting food application.

Acknowledgments

This research was supported by the Consejo Nacional de Ciencia y Tecnología (CONACYT) de México and the Programa de Mejoramiento al Profesorado (PROMEP) de la Secretaría de Educación Pública (SEP-México).

References

Alaiz, M.; Navarro, J.L.; Vioque, E.; Vioque, G. Amino acid analysis by high-performance liquid chromatography after derivatization with diethyl ethoxymethylenemalonate. *J. Chromatogr.* 1992, 591, 181-186.

Aliaga, C. and Lissi, E.A. Reactions of the radical cation derived from 2,2' azinobis(3-ethylbenzothiazoline-6-sulfonic acid) ($ABTS^{\cdot+}$) with amino acids. Kinetics and mechanism. *Canadian J. Chem.* 2000, 78(8), 1052-1059.

AOAC. Official Methods of Analysis, AOAC, Arlington, VA, USA. Secs. 920.39, 923.03, 925.09, 954.01, 962.09, 992.16, 1997.

Bahnassay, Y. and Khan, K. Fortification of spaghetti with edible legumes. II. Rheological, processing, and quality evaluation studies. *Cereal Chem.* 1986, 63(3), 216-219.

Bergman, C.; Gualberto, D.; Weber, C. Development of high-temperature-dried soft wheat supplemented with cowpea (*Vigna unguiculata* L. Walp): cooking quality, color and sensory evaluation. *Cereal Chem.* 1994, 71, 523-527.

Borneo, R. and Aguirre, A. Chemical composition, cooking quality, and consumer acceptance of pasta made with dried amaranth leaves flour. *LWT-Food Sci. and Tech.*, 2008, 41, 1748-1751.

Camire, M.E. Chemical and nutritional changes in food during extrusion. In M. N. Riaz (Ed.), Extruders in Food Applications (pp. 127-148). Boca Raton, FL: CRC Press. 2002.

Dahlin, K.M. and Lorenz, K.J. Carbohydrate digestibility of laboratory-extruded cereal grains. *Cereal Chem.* 1993, 70(3), 329-333.

Escudero Alvarez, E. and González Sánchez, P. La fibra dietética. *Nutr. Hosp.* 2006, 21(2), 61-72.

European Commision. Functional Foods. *European Commision/European Research Area/Food Agricultur & Fisheries & Biotechnology*, 2010, 1-28.

FAO/WHO. Protein quality evaluations. Report of joint FAO/WHO expert consultation. Food and nutrition. Paper No. 51. Rome, Italy: Food Agriculture Organization and the World Health Organization. 1991.

Hayakari, M.; Kondo, Y. A rapid and simple spectrophotometric assay of angiotensin-converting enzyme. *Anal. Biochem.* 1978, 84, 361-369.

Holm, J.; Björck, I.; Drews, A.; Asp, N.G. A rapid method for the analysis of starch. *Stärke/Starch.* 1986, 38, 224-226.

Hong, L.G.; Wei, L.G.; Liu, H.; Hui, S.Y. Mung-bean protein hydrolysates obtained with alcalase exhibit angiotensin I-converting enzyme inhibitory activity. *Food Sci. Technol. Int.* 2005, 11, 281-287.

Hurrel, R.; Lerman, P.; Carpenter, K. Reactive lysine in food stuff as measured by a rapid dye-binding procedure. *J. Food Sci.* 1979, 44, 1221-1231.

Iwaniak, A. Minkiewicz, P. Proteins as the source of physiologically and functionally active peptides. *Acta Sci Pol Technol Aliment*, 2007, 6(3):5-15.

Korhonen H., Pihlanto A. Technological options for the production of health-promoting proteins and peptides derived from milk and colostrum. *Curr. Pharm. Des.*, 2007, 13(8), 829-843.

Matsui, T.; Matsufuji, H.; Seki, E.; Osakima, K.; Nakashima, M.; Osajima, Y. Inhibition of angiotensin I-converting enzyme by Bacillus licheniformis alkaline protease hydrolyzates derived from sardine muscle. *Biosci Biotech Biochem*, 1993, 57:922

Meisel, H.; Bernard, H.; Fairweather-Tait, S.; FitzGerald, R.J.; Hartmann, R.; Lane, C.N.; McDonagh, D.; Teucher, B.; Wal, J.M. *Brit. J. Nutr.* 2001, 85:635.

Montgomery D. Diseño y análisis de experimentos. Limusa-Wiley, México, D.F., México, 2004.

Morales-De-León, J. C., Vázquez-Mata, N., Torres, N., Gil-Zenteno, L., and Bressani, R. Preparation and characterization of protein isolate from fresh and hardened beans (*Phaseolus vulgaris* L.). *J. Food Sci.*, 2007, 72, 96-102.

Nielsen, P.; Petersen, D.; Dammann, C. Improved method for determining food protein degree of hydrolysis. *J. Food Sci.* 2001, 66, 642-643.

Osorio, P.; Acevedo, E.; Carmona, R.; Tovar, J.; Paredes, O.; Bello, L. Resistant starch and *in vitro* starch digestibility of cooked "ayocote" bean (*Phaseolus coccineus*). *Rev Interciencia*. 2004, 29, 510-514.

Pedroche, J.; Yust, M.M.; Girón-Calle, J.; Alaiz, M.; Millán, F.; Vioque, J. Utilization of chickpea protein isolates for the production of peptides with angiotensin I-converting enzyme inhibitory activity. *J. Sci. Food Agric.* 2002, 82, 960-965.

Pérez, N. C. ; Cruz, E.R. ; Chel, G.L. ; Betancur, A.D. Effect of extrusion on nutritional quality of maize and Lima bean flour blends. *J. Sci. Food Agric.* 2006, 86(14), 2477-2484.

Prosky, L.; Asp, N.; Schweitzer, T.; Debris, S.; Furda, I. Determination of insoluble, soluble and total dietary fiber in food and food products: interlaboratory study. *J. Assoc. Offic. Anal. Chem.* 1998, 71, 1017-1023.

Pukalskas, A.; Van Beek, T.; Venskutonis, R.; Linssen, J.; Van Veldhuizen, A.; Groot, A. Identification of radical scavengers in sweet grass (*Hierochloe odorata*). *J. Agric. Food Chem.* 2002, 50, 2914-2919.

Rudloff, S. and Lönnerdal, B. Solubility and digestibility of milk proteins in infant formulas exposed to different heat treatments. *J. Pediatr. Gastroenterol Nutr.* 1992, 15, 25-33.

Ruiz-Ruiz, J.C.; Dávila-Ortíz, G.; Chel-Guerrero, L.A.; Betancur-Ancona, D.A. Wet fractionation of hard-to-cook bean (*Phaseolus vulgaris* L.) seeds and characterization of protein, starch and fiber fractions. *Food Bioproc. Technol.* 2012, 5(5), 1531-1540.

Seppo L., Kerojoki, O., Suomalainen, T., Korpela, R. The effect of a Lactobacillus helveticus LBK-16H fermented milk on hypertension- a pilot study on humans. *Milchwissenchaft*, 2002, 57:124.

Shimizu, T. Health claims on functional foods: the Japanese regulations and an international comparison. *Nutr. Res. Rev.* 2003, 16(2): 241-252.

Shogren, R.L.; Hareland, Y.; Wu, Y. Sensory evaluation and composition of spaghetti fortified with soy flour. *J. Food Sci.* 2006, 71, S428-S432.

Tovar, J.; Bjorck, I.; Asp, N.G. Starch content and alpha-amylolysis rate in precooked legume flours. *J. Agric. Food Chem.* 1990, 38, 1818-1823.

Udenigwe, C.C. and Aluko, R.E. Food protein-derived bioactive peptides: production, processing and potential health benefits. *J. Food Sci.* 2012, 71(1): R11-R24.

Yokoyama, K.;Chiba, H.; Yoshikawa, M. Peptide inhibitors for angiotensin I-converting enzyme from thermolysin digest of dried bonito. *Biosci Biotech Biochem*, 1992, 56:1541.

Zamora, N.C. Efecto de la extrusión sobre la actividad de factores antinutricionales y digestibilidad *in vitro* de proteínas y almidón en harinas de *Canavalia ensiformis*. *Arch. Latin Nutr*. 2003, 53 (3), 293-298.

In: Processed Foods
Editors: Chloe M. Gagne and Daniel B. Jones

ISBN: 978-1-62808-688-1

Chapter 7

BASIC GUIDELINES FOR ESTABLISHING FOOD STANDARDS

Wenbiao Wu*
College of Food Science, Southwest University, Tian Sheng Qiao, Chongqing, People's Republic of China

ABSTRACT

Food manufacturers need a previously established standard to produce high quality food products with uniformity, which is always concerned by consumers. This article aims to introduce the basic guidelines that should be followed in developing a common accepted food (especially new food) standard. Based on the literature review and evaluation of the current status of food quality standards, the basic guidelines for establishing a food standard have been developed. The contents of this guideline include *Determining the critical interior characteristics of a food product required by its standard, Determining the quality requirements of raw materials and additives for producing a food product by its standard, Determining the critical foreign material contamination of a food product required by its standard, Determining the required processing, storing or selling conditions of a food product by its standard, Determining the critical limit level of the required characteristics of the food product by its standard, Establishing the framework of the standard* and *Drafting the standard according to the framework established.* The publication of this guideline should be able to help any company or agency to establish a well common accepted standard for a well established or newly developed food product.

Keywords: Basic guidelines, Food standard, Quality, Normative references, Control

* Correspondent: wbwu62@yahoo.com.

INTRODUCTION

Although, standards could be divided into process, product or information (Caswell, 2003), most regulations use a combination of the three to regulate food processing and marketing (Vieira, 2006). Food standards may be classed as the assignment of the properties or characteristics of a product as well as the processing, storing or selling conditions for maintaining these required properties, which are ratified by an appropriate governmental agency. The properties or characteristics of a food product include nutrients, hazards, sensory properties and package specifications. Nutrients include amino acids, proteins, saccharides, fatty acids, oils or fats, essential nutritional minerals and vitamins. Hazards include toxic inorganic and organic substances, pathogenic microorganisms. Sensory properties include taste, color, smell, texture, form and morpha. Package property is related to quantity, size, materials and strength. The exact context of the standard for a food product includes the assignment of the exact values of these characteristics, such as protein or Pb content. The standards for the sensory properties and some aspects of package commonly represent the consumer's requirements for a product. Sensory properties are vital for a food product having sales ability. If the sensory properties are included in a food standard that is mandatory, they must be met. The processing, storing or selling conditions may include the methods of handling or processing, temperatures, times, air condition, pH, etc.

Food standards may be drafted by a governmental agency, an organization, a food company or any other food related enterprise, etc. Therefore, food standards can be grouped into legal standards, company or voluntary label standards, industry standards and consumer or grade standards. Legal standards of a food product are usually related to its nutrient or hazard substance content and packing quantity, which is established by the federal, state or municipal agencies. No matter who is going to establish them, it is legally mandatory that they must be ratified by an appropriate governmental agency throughout the whole world.

The principle aim of establishing a food standard by food companies or other food related enterprises is to assure the acceptability of their products to the consumers. The legal aim of establishing a food standard before producing and selling the food is to protect consumer's benefit. Furthermore, nowadays, food standards have become mandatory requirements for international trade (Donovan et al, 2001; Henson & Loader, 2001; Farina & Reardon, 2000).

Although efforts have been made to promote internationally harmonized standard for the same well established food product (Randell, 1991; Dawson, 1995; Jukes, 2000; Fehér, 2002; Boutrif, 2003), there are still many documented and published standards established by different agencies or companies that are widely used in the food market (Reardon et al., 2001; Reardon & Farina, 2001). For example, Ouma (2010) stated: "Over the past decade, private food safety and quality standards have become focal points in the supply chain management of large retailers, reshaping governance patterns in global agrifood chains.". Furthermore, for the development of a new food product by food technologists or its production by manufacturers, it is essential to establish a new standard. It will help to avoid the confusion and disruption that multiple audits can bring. This could not be found in past literatures. Therefore, this article is to introduce the basic guidelines for developing standards applicable to well established or newly developed food products.

DETERMINING THE CRITICAL INTERIOR CHARACTERISTICS OF A FOOD PRODUCT REQUIRED BY ITS STANDARD

The determination of the critical interior characterists of a food product which should be required by its standard is very important and difficult. It should be estimated very carefully. It is quite difficult to find any standard guideline for how to determine the critical interior characterists of a food product which should be required by its standard in the literature. However, there are three aspects that should be taken into consideration. Consumer's concern is the one that should firstly be met, which may be evaluated by a market research such as consumer's acceptability study. Secondly, scientific interventions such as by nutritional and dietary safety researching results aimed at the modification of consumer's food choices should be considered (Fitzgerald et al., 2010). Thirdly, the nature of raw materials or main recipe components may also be a key factor for determining the interior characterists control index of a food product which should be required by its standard. For example, other nutritional components in potato crisps may not be necessary to be included in its interior characteristics control index while its fat content and water content are always required to be accurately controlled below appropriate values in its standard. On the other hand, some other foods, such as artificial milky powder recipes for infants should contain appropriate amount of every essential nutrient, which is always required by any kind of its standard. The three guidelines mentioned above are also applicable to the assignment of sensory properties in a food standard, etc. For instance, potato crisps shall have a light yellow to golden brown colour and crispy texture, which are always required in its standard only because they are liked by consumers.

DETERMINING THE QUALITY REQUIREMENTS OF RAW MATERIALS AND ADDITIVES FOR PRODUCING A FOOD PRODUCT BY ITS STANDARD

The quality requirements of raw materials and additives for producing a food product by its standard are mainly related to dietary safety considerations. Generally, the dietary safety of any raw material or additive that are going to be used for producing a food must be approved by sufficient scientific evidences and related authorities. Otherwise, they should be already and safely taken as foods by humans for long terms, such as potatoes, rice, meat, sweet potatoes. These requirements may be optional and might not be necessary to be assigned in some food standards, such as the characteristics of the potato tubers that are used for producing potato flakes are normally not required in the potato flakes standard. However, the quality requirements of raw materials and additives might be assigned in some food standards because of technical considerations on quality assurance.

Determining the Critical Foreign Material Contamination of a Food Product Required by Its Standard

The critical foreign material contamination of a food product usually include toxic minerals or organic substances, pathogenic microorganisms, pesticide or herbicide residues and other unwanted materials that can negatively affect consumer's choice. The requirements of limiting these materials contents for producing a food product by its standard are mainly related to dietary safety considerations. It is quite difficult to find any standard guideline for how to determine the critical foreign material contaminants of a food product which should be required by its standard in the literature. However, the guidelines for determining the critical foreign material contaminants which could be followed may include: finding any information about or measuring the variety of toxic substances and their contents presented in soils at where the raw materials used for making the food are planted or irrigating water, analyzing any possibility of contaminants from food processing equipment or package materials or air or operating persons, understanding or investigating the culturing measures such as pest or herb control of the raw materials, considering the nature of the food being processed.

Determining the Required Processing, Storing or Selling Conditions of a Food Product by Its Standard

The required processing, storing or selling conditions of a food product by its standard are mainly related to technical considerations on quality assurance. These requirements may be optional and might not be necessary to be assigned in some food standards. The required processing, storing or selling conditions of a food product by its standard are usually determined by appropriate scientific studies. The determination of the critical processing, storing or selling conditions of a food product which should be required by its standard is also very important and difficult. It could firstly be estimated by referring to published information in the literature. If this still can not solve the problem, appropriate experimental studies are necessary.

Determining the Critical Limit Level of the Required Characteristics of the Food Product by Its Standard

For a well established food product, the critical limit level of the required characteristics that may include nutrients, hazards, sensory properties and package specifications could be found in the past normative references. The quotations from the past normative references should be good enough to make good quality food product with uniformity.

For a new food product (especially produced with a new raw food materials or the addition of a new ingredient), the critical limit level of the required characteristics that may include nutrients, hazards, sensory properties and package specifications should be determined by scientific researches. For example, the dietary safety or toxicity /or the critical limit level of hazards should be evaluated /or determined by following OCED or other appropriate guidelines. The consumer acceptability study should be carried out to found out the critical limit of sensory properties. Experimental studies should also be necessary for determining the required nutritional and package specifications. For example, usually, the fat contents of fat and water in potato crisps are <30% or 40% or 45%, and <5% or 3%, respectively, which are normally determined by consumer acceptability study and laboratory test results of being able to assure crispy texture, etc.

ESTABLISHING THE FRAMEWORK OF THE STANDARD

Establishing the framework is the first step that should be taken in establishing a food standard. This is very important for different agencies or companies from different countries or regions to make a common accepted food standard.

World Standard Services Network (www.wssn.net) provides the list of international, regional and national standardization agencies or organizations and their related activities or services through its web site. These agencies or organizations usually provide template or format for structuring a standard document. The organizations that have great impact on the world include FAO/WHO, World Trade Organization (WTO), International Organization for Standardization (ISO), Organization for Economic Cooperation and Development (OECD), Euroasian Interstate Council for Standardization, Metrology and Certification. If the standard being drafted is going to apply for any one of these organization's standards, its format or template should be the same as that provided by them. For example, if the standard being drafted is going to apply for FAO/WHO Codex Alimentarius, its format or template should be the same as that provided by them. If the standard being drafted is going to apply for ISO standard, its format or template should be the same as that provided by this organization. For the whole format and the way to write the content of the commodity standard under the title of each section in detail, ISO (2002, 2004 and 2007) can be referred to.

For establishing many other food standards such as private standards being established by food companies themselves, the proposed format outline of a food standard that is shown in Table 1 could be used. (This framework has been developed on the basis of the food standard formats proposed by organizations mentioned above and is more complete than others. This table may help any agency or company to establish a well common accepted standard for a food product.)

Drafting the Standard According to the Framework Established

After the framework is established (see Table 1), this step can be taken in drafting the food standard. Here, the statements that should be included in the food standard as appropriate under the relevant headings of the standard are described.

Giving the Heading of the Standard

The heading of the food standard should be as common, clear and concise as possible. If the food standard attempts to cover more than one food, its name should be generic and able to cover them all. A subtitle could be added in the case of a fully informative title being too long. The heading of the standard may be given like that is shown in Appendix A.

Table 1. Suitable format or template for self-structuring a food standard by a food company or enterprise

<Name of the food> National (or Industry or Enterprise, etc) Standard	
1 Scope of the standard 2 Delineation of normative references and terms 2.1 Delineation of normative references 2.2 Delineation of terms 2.2.1… … 3 Raw materials and Additives quality requirements 3.1 Raw materials 3.1.1… … 3.2 Additives 3.2.1… … 4 Product interior characteristics control index 4.1 Chemical and physical characteristics 4.1.1 Chemical characteristics 4.1.1.1… … 4.1.2 Physical characteristics 4.1.2.1… … 4.2 Sensory characteristics 4.2.1… … 5 Foreign material contamination control index 5.1 Pesticide residues 5.1.1 … … 5.2 Toxic minerals 5.2.1 … … 5.3 Pathogenic microorganisms 5.3.1 … … 5.4 … …	6 Packing, transporting and storing requirements 6.1 Packing requirements 6.1.1 Container 6.1.2 Packing forms 6.1.3 Weights (or Volumes or Sizes) 6.1.4 Packing conditions 6.1.5 Labelling 6.1.6 Marking … 6.2 Transporting requirements 6.2.1 … … 6.3 Storing requirements 6.3.1 … … 7 Product quality testing 7.1 Testing procedures 7.1.1… … 7.2 Testing forms 7.2.1 Test for product sending 7.2.2 Type test 7.2.3 Batch test 7.2.4 Sampling test 7.3 Decision rules 7.3.1 … … 8 Recall Annotations A. Drafters: B. Submitter: C. Submitting date: D. Ratifier: E. Ratifying date: Appendixes Bibliography

Editing the "1 Scope of the Standard"

The "*1 Scope of the standard*" should summarize what the standard document does. It should be written as a series of statements of fact. It should also state clearly and concisely which food or foods the standard is applicable to.

Editing the "2 Delineation of Normative References and Terms"

The text of the "*2.1 Delineation of normative references*" could be fixed. It could be written like the form that is shown in Appendix A.

For the "*2.2 Delineation of terms*", it may be given in the normative references. In this case, it is only necessary to cite these references. There may also be additional terms required to clarify the meaning of the standard. The self-explanatory of these terms should be given under this heading. This section of the food product standard draft may take the form that is shown in Appendix A.

Editing "3 Raw Materials and Additives Requirements"

This heading covers two subtitles including "*3.1 Raw materials requirements* and *3.2 Additives requirements*". All raw materials and additives permitted to produce the food product should be listed and clearly described under each subtitle, respectively. Any necessary normative references to types, or styles, or quality including dietary safety or maximum amount allowed of the raw materials and additives in the food should also be quoted. If there is no normative reference, any published report in the literatures or testing report by authority certified bodies related to the above mentioned aspects of the raw materials and additives in the food may be quoted. This section of the food product standard draft may take the form that is shown in Appendix A.

Editing "4 Product Interior Characteristics Control Index"

This heading covers two subtitles including "*4.1 Chemical and physical characteristics*" and "*4.2 Sensory characteristics*". Under the heading - "*4.1 Chemical and physical characteristics*", it should be subdivided into "*4.1.1 Chemical characteristics*" and "4.1.2 *Physical characteristics*". For "*4.2 Sensory characteristics*", it may not be necessarily subdivided.

The context of "*4.1.1 Chemical characteristics*" should give all quantitative and other requirements as to the chemical composition such as protein content, or other nutrients content, or other non nutritional components (including endogenous toxin) content of the food product concerned. They are main parts of the quality factors which are essential for the designation or definition of the food product concerned.

The context of "4.1.2 *Physical characteristics*" should give all requirements as to the physical properties such as shapes, forms or sizes of the food product concerned. They may be parts of the quality factors which are essential for the designation or definition of the food product concerned.

The context of *"4.2 Sensory characteristics"* should give all requirements as to the sensory properties such as taste, odour, colour and texture of the food product concerned. They may also be the basic quality criteria for the finished food product concerned.

Under each heading, any necessary normative references to critical quality factor related to chemical and physical composition or sensory properties of the food product concerned should also be quoted. If there is no normative reference, any published report in the literatures or testing report by authority certified bodies related to the above mentioned aspects of the food product may be quoted.

Editing "5 Foreign Material Contamination Control Index"

This heading may cover four subtitles including "*5.1* Pesticide residues*", "5.2 Toxic minerals*", "*5.3 Pathogenic microorganisms*" and *"5.3 Other foreign contaminants"*. Under each heading, all foreign materials or substances permitted to exist in the food product concerned should be listed and clearly described, respectively. Any necessary normative references to the maximum amount allowed of these foreign materials or substances in the food product concerned should also be quoted. If there is no normative reference, any published report in the literatures or testing report by authority certified bodies related to the above mentioned foreign materials or substances in the food product concerned may be quoted. This section of the food product standard draft may take the form that is shown in Appendix A.

Editing "6 Packing, Transporting and Storing Requirements"

This heading covers three subtitles including "*6.1 Packing requirements*", "*6.2 Transporting requirements*" and "*6.3 Storing requirements*". Under the heading - "*6.1 Packing requirements",* it should be subdivided into "*6.1.1 Container", "6.1.2 Packing forms", "6.1.3 Weights (or Volumes or Sizes)*", "*6.1.4 Packing conditions*", "*6.1.5 Labelling*", "*6.1.6 Marking*", etc.. For "*6.2 Transporting requirements"* and *"6.3 Storing requirements"*, they may not be necessarily subdivided and these two clauses are optional.

Under "*6.1.1 Container",* the requirements on chemical, physical and sensory properties as well as constituent materials of the container should be assigned in addition to the citation of any existing normative references.

Under "*6.1.2 Packing forms",* the preferred packing forms should be assigned in addition to the citation of any existing normative references.

Under *"6.1.3 Weights (or Volumes or Sizes)*", the preferred packing weights (or volumes or sizes) should be assigned in addition to the citation of any existing normative references.

Under "*6.1.4 Packing conditions*",

Under "*6.1.5 Labelling*", any information that should appear on the label should be stated clearly in addition to the citation of any existing normative references.

Under "*6.1.6 Marking*", it should be stated clearly how the package shall be marked in addition to the citation of any existing normative references.

Under each heading, any necessary normative references to any packing, transporting or storing requirements of the food product concerned should also be quoted. If there is no normative reference, any published report in the literatures or testing report by authority certified bodies related to the above mentioned aspects in the food product concerned may be quoted. This section of the food product standard draft may take the form that is shown in Appendix A.

Editing "7 Product Quality Testing"

This heading covers three subtitles including "*7.1 Testing procedures*", "7.2 *Testing forms*" and "*7.3 Decision rules*".

Under "*7.1 Testing procedures*", it may be necessarily subdivided into many clauses so that the test of all critical quality factors can be covered. Generally, only normative references should be quoted. If there is no normative reference, any published report in the literatures or testing report by authority certified bodies related to the above mentioned aspects in the food product concerned may be quoted.

Under "7.2 *Testing forms*", it is subdivided into "*7.2.1 Test for product sending*", "*7.2.2 Type test*", "*7.2.3 Batch test*" and "*7.2.4 Sampling test*". Under "*7.2.1 Test for product sending*", it should be clearly stated what critical quality factors must be tested and who should do this before the food product concerned in factory is sent to market for selling. Under "*7.2.2 Type test*", the exact time interval of testing for regular inspection and the critical factor items that need to be tested should be clearly stated. In addition, it should also be stated clearly that there is a need to carry out a "*Type test*" in the case of what happens. Under "*7.2.3 Batch test*", the definition of this test and the critical factor items that need to be tested should be clearly stated. Under "*7.2.4 Sampling test*", the place and number of samples should be clearly stated. The number of samples for sensory, chemical and physical, or pathogenic microorganism tests should also be clearly stated, respectively. Under "*7.3 Decision rules*", the requirements that should be met by testing results for judging the food product that is qualified should be cleared stated. This section of the food product standard draft may take the form that is shown in Appendix A.

Editing "8 Recall"

The food product that is sold in the market should be recalled in case of what happens by who should be clearly stated. This section of the food product standard draft may take the form that is shown in Appendix A.

Additional General Guidelines

According to the variety of the food product, the clauses listed in Table 1 and described above should be selected properly. Some of them may be omitted depending upon the nature or characteristics of the food product concerned.

Table 2. Comparison between the frameworks of a food standard proposed by FAO/WHO Codex Alimentarius Commission (2007), ISO (2007) and this article

This article	FAO/WHO	ISO
Heading of the standard	Name of the standard	Title of the standard
1 Scope of the standard	Scope	1 Scope
2 Delineation of normative references and terms	Description	2 Normative references
3 Raw materials and Additives quality requirements	Essential composition and quality factors	3 Terms and definitions
4 Product interior characteristics control index	Food additives	4 Requirements
5 Foreign material contamination control index	Contaminants	5 Sampling
6 Packing, transporting and storing requirements	Hygiene	6 Test methods
7 Product quality testing	Weights and measures	7 Test report
8 Recall	Labelling	8 Packaging
Annotations	Methods of analysis and sampling	9 Marking
Appendixes		Annexes
Bibliography		Bibliography

Appendix A. An Introductive Sample of a Food Standard Draft

< Note: the name of the food, country, company, etc -This phrase should not appear in the context of drafted standard >* National (or Industry or Enterprise, etc) Standard,

e.g. Potato Crisps Chinese National Standard.

1. Scope of the Standard

This standard assigns minimum requirements and methods of sampling and testing for potato crisps.

It is applicable to potato crisps (American name; potato chips) made from fresh potato tubers by frying.

2. Delineation of Normative References and Terms

2.1 Delineation of normative references

The clauses in the referenced normative documents become parts of the provisions under any relevant headings of this standard document. For dated normative documents that referenced, only the edition cited applies. All revised editions or amendments of the dated normative documents that referenced do not apply accept the statements of errata list. However, all parties who are using this standard are encouraged to negotiate with each other about the acceptance of the latest edition of the referenced normative documents. For undated normative documents that referenced, the latest edition including any amendments applies.

GB5009.6, Testing method of fat in foods

GB2713, Hygienic indexes of starch product

...

2.2 Delineation of terms

This standard adopts the following terms.

<Note: Then, the description of terms required should be given one by one. -This phrase should not appear in the context of drafted standard>*

Potato crisps: Deep-fried crispy thin slices of washed and peeled fresh potato tubers.

...

3. Raw Materials and Additives Requirements

3.1 Raw materials requirements

For this standard, < Note: the name of the normative references or other appropriate documents-This phrase should not appear in the context of drafted standard >* ZB B23008□85 is quoted.

< Note: Then should follow a list and description of the raw materials, e.g.: "Name of raw material, level (in percentage)" -This phrase should not appear in the context of drafted standard >*

Fresh potato tubers, 100%

3.2 Additives requirements

For this standard, < Note: the name of the normative references or other appropriate documents-This phrase should not appear in the context of drafted standard >* GB/T 18770-2002 is quoted.

< Note: Then should follow a list and description of the additives, e.g.: "Name of additive, maximum level (in percentage)" -This phrase should not appear in the context of drafted standard >*

Salt (NaCL), <1.5%

4. Product Interior Characteristics Control Index

4.1 Chemical and physical characteristics

4.1.1 Chemical characteristics

Fat content: < 30%

Water content: <3%

4.1.2 Physical characteristics

Thickness: <0.85-1.5 mm

4.2 Sensory characteristics

Color: a light yellow to golden brown colour

Texture: crispy

…

5. Foreign Material Contamination Control Index

5.1 Pesticide residues

For this standard, < Note: the name of the normative references or other appropriate documents-This phrase should not appear in the context of drafted standard.> GB 25193—2010 is quoted.

<Note: Then should follow a list and description of the pesticide, e.g.: "Name of pesticide, maximum level (in percentage or mg/kg)".>[*]

Chlorothalonil: < 5 mg/kg

Chlorpropham: < 30 mg/kg

…

5.2 Toxic minerals

For this standard, <Note: the name of the normative references or other appropriate documents>□EC□NO. 629/2008 is quoted.

<Note: Then should follow a list and description of the toxic minerals, e.g.: "Name of toxic mineral, maximum level (in percentage or mg/kg)".>[*]

Pb: < 3 ppm

Hg: < 0.1 ppm

…

5.3 Pathogenic microorganisms

For this standard, <Note: the name of the normative references or other appropriate documents> GB2713 is quoted.

<Note: Then should follow a list and description of the pathogenic microorganism, e.g.: "Name of pathogenic microorganism, maximum level (in cfu/ml or cfu/g)".>[*]

Total viable count, CFU per gram, max□500

Escherichia coli, CFU per gram: < 30

Pathogens: not detected

…

5.4 Other foreign contaminants

For this standard, <Note: the name of the normative references or other appropriate documents>[*] GBT 12729.5-91 is quoted.

< Note: Then should follow a list and description of other foreign contaminants, e.g.: "Name of foreign contaminant, maximum level (in percentage or mg/kg)".>[*]

Hairs: absent

…

6. Packing, Transporting and Storing Requirements

6.1 Packing requirements

6.1.1 Container

For this standard, <the name of the normative references or other appropriate documents>[*] GB 6544-1999 is quoted.

The container must not transmit any odour or flavour to the food product and must not contain substances that may damage the food product or constitute a human health risk. The container must be well (or properly) sealed without any leakage (or with proper ventilation, etc.).

The container is made from fiberboard.

6.1.2 Packing forms

The packing forms of the food product are inflatable plastic bags.

6.1.3 Weights <or Volumes or Sizes>

For this standard, <the name of the normative references or other appropriate documents>* JJG564-2002 is quoted.

Packing weight: 50±1.5g, … <x ± y (g, or ml, or kg, etc.). [or Packing size: X x Y x Z (mm, or cm, etc.)]>*

6.1.4 Packing conditions

For this standard, <the name of the normative references or other appropriate documents>* GB9681-9693 is quoted.

This food product should be packed at the following conditions: low humidity, no harmful gas, ...

6.1.5 Labelling

For this standard, <the name of the normative references or other appropriate documents>* GB 7718-2004 is quoted.

The label should indicate the following items clearly: name of the food product, the name and address of manufacturer, product specifications, quantity, transporting and storing instructions, application, usage, lot identification, traceability, etc.

6.1.6 Marking

For this standard, <the name of the normative references or other appropriate documents>* GB 7718-2004 is quoted.

The mark of packages shall conform to that required by the country of destination.

6.2 Transporting requirements

This food product shall never be pressed by heavy weight and shall be transported at cool and dry conditions. …

6.3 Storing requirements

This food product shall never be pressed by heavy weight and shall be stored at cool and dry places. …

7. Product Quality Testing

7.1 Testing procedures

7.1.1 Determination of fat content

For this standard, <the name of the normative references or other appropriate documents> GB5009.6 is quoted.

7.1.2 Determination of water

For this standard, <the name of the normative references or other appropriate documents> GB5009.3 is (or are) quoted.

7.1.3 …

…

7.2 Testing forms

7.2.1 Test for product sending

7.2.1.1 Before every batch of the food product being sent to market for selling, it must be improved in conformity with this standard through the test carried out by quality control group.

7.2.1.2 The testing items for product sending include pathogenic microorganisms, moisture content, net weight, …

7.2.2 Type test

7.2.1 The *Type test* shall be undertaken every … (e.g.6) mouths for regular inspection. It should be undertaken in case of the following: a) changing raw materials, b) seasonal production or restarting after stopped production for a long time, c) restarting after changed production equipment, d) required by … (e.g. national product supervision administration)

7.2.2 The items of *Type test* include all assigned in this standard.

7.2.3 Batch test

The food products with the same specification manufactured by the same shift, production line, and raw materials defined as the same batch.

7.2.4 Sampling test

7.2.4.1 The sample shall be collected from product store. The sampling unit shall be … (e.g. box or package)

7.2.4.2 Sampling rate for every batch shall be … (e.g. 0.3%). Total sampling numbers for every batch shall not less than … (e.g. 10 bales) with total quantities shall not be less than … (e.g. 3 kg). … (e.g. 3 bales) of these samples are for sensory and net weight test. … (e.g. 3 bales) of these samples are for chemical and physical test. … (e.g. 3 bales) of these samples are for pathogenic microorganism test.

7.3 Decision rules

7.3.1 Decision rules for Test product sending

7.3.1.1 The food product with that all testing items meet the requirements assigned in this standard is qualified.

7.3.1.2 If the food product with that one item does not meet the requirements assigned in this standard accept pathogenic microorganisms, it may be re-tested through double sampling. If the result of re-testing proves that this item does not still meet the requirements appointed in this standard, the food product is not qualified.

7.3.1.3 The food product with that the testing of even only pathogenic microorganism does not meet the requirements assigned in this standard is not qualified. Re-testing should not be undertaken.

7.3.2 Decision rules for Type Test

7.3.2.1 The food product with that all testing items meet the requirements assigned in this standard is qualified.

7.3.2.2 If the food product with that less than three items do not meet the requirements assigned in this standard accept pathogenic microorganisms, it may be re-tested through double sampling. If the result of re-testing proves that one item remains to be not in conformity with the requirements assigned in this standard, the food product is not qualified. The food product with that more than three items do not meet the requirements appointed in this standard is not qualified and re-testing should not be undertaken.

7.3.2.3 The food product with that the testing of even only pathogenic microorganism does not meet the requirements assigned in this standard is not qualified. Re-testing should not be undertaken.

8. Recall

If the storing time of the food product being sold in the market is over the shelf life negotiated by manufacturer and retailer or its quality is not satisfied, it should be recalled.

*: *Words or phrases inside '< >' shall never appeared in the finished food standard draft.*

REFERENCES

Boutrif, E. (2003). The new role of Codex Alimentarius in the context of WTO/SPS agreement. *Food Control*, 14 (2), 81-88.

Caswell, J. A. (2003). Food Safety. In L. J. Unnevehr (Ed.), Food Safety in Food Security and Food Trade. Washington: International Food Policy Research Institute.

Dawson, R.J. (1995). The role of the Codex Alimentarius Commission in setting food standards and the SPS agreement implementation. *Food Control*, 6(5), 261-265.

Donovan, J. A., Caswell, J. A. & Salay, E. (2001). The effect of stricter foreign regulations on food safety levels in developing countries: a study of Brazil. *Review of Agricultural Economics*, 23(1), 163-175.

FAO/WHO (2007). Joint FAO/WHO Food Standard Programme, Codex Alimentarius Commission, Procedural Manual, 17th edition, Pp. 100-103. 00153 Rome, Italy.

Farina, E. M. M. Q. & Reardon, T. (2000). Agrifood grades and standards in the extended Mercosur: their role in the changing agrifood system. *American Journal of Agricultural Economic*, 82(5), 1170-1176.

Fehér, I. (2002). Meeting EU standards in Eastern Europe: the case of the Hungarian agri-food sector. *Food Control,* 13 (2), 93-96.

Fitzgerald, A., Heary, C., Nixon, E. & Kelly, C. (2010). Factors influencing the food choices of Irish children and adolescents: a qualitative investigation. *Health Promotion International*, 25(3), 289-298. doi:10.1093/heapro/daq021

Henson, S. & Loader, R. (2001). Barriers to agricultural exports from developing countries: the role of sanitary and phytosanitary requirements. *World Development,* 29(1), 85-102.

ISO (2002). Basic template of drafting an international standard. ISO/TC 000/SC 0 N 000.

ISO (2004). ISO/IEC Directives, Part 2. Rules for the Structure and Drafting of International Standard (5th edition).

ISO (2007). Model document of a draft International Standard,. Rice model/Version 2.1/2007-08-15.

Jukes, D. (2000). The role of science in international food standards. *Food Control*, 11 (3), 181-194.

Ouma, S. (2010). Global Standards, Local Realities: Private Agrifood Governance and the Restructuring of the Kenyan Horticulture Industry. *Economic Geography,* 86(2), 197-222.

Randell, A.W. (1991). FAO/WHO conference on food standards, chemicals in food and food trade. *Food Control*, 2 (Issue 3), 185-186.

Readon, T., Codron, J.M., Busch, L., Bingen, J. & Harris, C. (2001). Global changes in agrifood grades and standards: agribusiness strategic responses in developing countries. *International Food and Agribusiness Management Review*, 2(3-4), 421-435.

Readon, T. & Farina, E.M.M.Q. (2001). The rise of private food quality and safety standards: illustrations from Brazil. *International Food and Agribusiness Management Review*, 4(4), 413-421.

Vieira, L.M. (2006). The Role of Food Standards in International Trade: Assessing the Brazilian Beef Chain. *Brazilian Administration Review*, 3(1), Art. 2, 17-30.

In: Processed Foods
Editors: Chloe M. Gagne and Daniel B. Jones

ISBN: 978-1-62808-688-1

Chapter 8

STRATEGIES FOR ANNATTO SEEDS PROCESSING WITH PRESSURIZED FLUIDS IN FOOD INDUSTRIES

D. T. Santos, L. M. Rodrigues, R. A. C. Torres, M. T. M. S. Gomes and M. A. A. Meireles*

LASEFI (Laboratory of Supercritical Technology: Extraction, Fractionation, and Identification of Vegetable Extracts) / DEA (Department of Food Engineering) / FEA (School of Food Engineering) / UNICAMP (University of Campinas), R. Monteiro Lobato, Campinas, São Paulo, Brazil

ABSTRACT

Nowadays, food scientists have collaborated with nutrition researchers to develop plant-based food products to promote healthy eating habits. In food research, carotenoids from fruits and vegetables have attracted a great deal of attention for their functional properties, health benefits and prevention of several major chronic diseases. Annatto (*Bixan orellana L.*) is a tropical tree whose seeds produce carotenoid pigments, oil-soluble bixin and water-soluble norbixin, which have a widespread use in the food industry, namely for coloring butter, margarine, cheese, oils and sauces, with hues ranging from yellow to red. Carotenoids are unstable compounds, very sensitive to oxidation processes. To avoid these effects some caution during its processing should be taken into consideration. The processing of carotenoids with pressurized fluids in liquid, subcritical or supercritical state at selected conditions has demonstrated to be a promising option. This chapter presents the recent works done and under evaluation by our research group (LASEFI/DEA/FEA/UNICAMP) for obtaining annatto seeds products: extracts, coprecipitates, nanosuspensions and nanoemulsions aiming higher yields, low manufactured cost with concomitant minimal carotenoid degradation.

* E-mail address: meireles@fea.unicamp.br.

INTRODUCTION

Annatto (*Bixa orellana* L.) is a tropical tree whose seeds produce pigments which have a widespread use in the food industry, namely for colouring butter, margarine, cheese, oils and sauces, with hues ranging from yellow to red. The pigments are located in the pericarp of the seeds and have two main components: oil-soluble bixin and water-soluble norbixin [1].

It has the natural colorant bixin, which is the most widely used colorant in food processing [2]. The pigments of the seeds, bixin and norbixin, are amongst those most used in the food, pharmacological and cosmetic industries due to the intensity of their colors and the wide variety of tones from yellow to red. This range of colors is an additional advantage of the annatto carotenoids over other ones, such as those of the carrot and beetroot, which only show their respective colors [3].

More than 80% of the total pigments in the seeds coat of annatto consists of the carotenoid bixin, 6-methyl hydrogen 9'-cis-6,6'-diapocarotene-6,6'-dioate [4]. The long system of conjugated double bonds present in the structure of bixin is responsible for the red color, but also for the poor stability of the molecule under processing and storage conditions such as high temperature, light and oxygen [5].

Nowadays, food scientists have collaborated with nutrition researchers to develop plant-based food products to promote healthy eating habits. In food research, carotenoids from fruits and vegetables have attracted a great deal of attention, mainly focused on the analysis of geometric carotenoid isomers. Carotenoids found in fruits and vegetables have also attracted great attention for their functional properties, health benefits and prevention of several major chronic diseases [6].

The carotenoid composition in different parts of plants is variable and made complex by low levels of biosynthetic precursors and derivatives of the main components. Animals are incapable of biosynthesizing carotenoids, thus depending on dietary carotenoids intake, which are selectively or not selectively absorbed, converted to vitamin A, deposited as such or slightly altered to form carotenoids typical of animal species. Consequently, carotenoids are not as widely distributed in animal-derived foods and are present at much lower levels [7].

An important method applied for the commercial extraction of annatto pigments from the *B. orellana* fruit consists of heating the suspension of seeds in oil to a maximum temperature of 130 °C under vacuum [8]. It is important to mention that the temperature appears as one of the main sources of the bixin molecule degradation. Near this extraction temperature, bixin undergoes a series of degradation reactions. The colorant degradation is observable by color changes, which is an important characteristic for the food industry as by controlling the temperature and duration of the heating process the red/yellow balance of this colorant can be tuned [9].

The possibility of tailoring the colorant end color opens a vast range of use. Nevertheless, the degradation or structural modification of the bixin is accompanied by the release of *m*-xylene, toluene, toluic acid, and toluic acid methyl ester, all of which are undesirable in preparations intended for food use [10]. Extraction with organic solvents and extraction into aqueous alkali are also important commercial processes used to extract the pigment from dried annatto seeds.

In particular, the stability of bixin was evaluated under the effect of several factors, such as light, oxygen concentration, temperature, and solvent [11]. Bixin showed great stability in the dark, in the presence or absence of oxygen, but under direct illumination it was degraded. The thermo- and photostability of the 9'-cis double bond in methyl bixin was reported as being so great that the molecule underwent a second trans-cis isomerization given considerable amounts of the 9, 9'-di-cis isomer, but in the presence of iodine and light, 9'cis-trans isomerization occurred [12]. Therefore, it seems that the fade of bixin could be strongly dependent on the experimental conditions (e.g., direct illumination, heating, photosensitization, etc.) [13].

Commonly, conventional extraction methods are used to extract carotenoids from plant sources. Nevertheless, these methods are, in general, time- and solvent-consuming and may promote the degradation of these compounds. On the other hand, bixin-rich extracts can undergo a series of degradation reactions during the post-production process and/or while stored if exposed to high temperatures, light and air, which results in restricted commercial applications. One of the alternatives that can be used to improve bixin stability is its encapsulation. Encapsulation has become an important tool, helping to increase shelf life and protecting the biological properties of the material [14]. Moreover encapsulation allows the controlled release of these functional pigments under desired conditions, for instance when ingested in the body [15].

In this context, this chapter presents the recent works done and under evaluation by our research group (LASEFI/DEA/FEA/UNICAMP) for obtaining annatto seeds products containing bixin: extracts, coprecipitates, nanosuspensions and nanoemulsions aiming higher yields, low manufactured cost with concomitant minimal carotenoid degradation. We are focused on the use of pressurized fluids in liquid, subcritical or supercritical state at selected conditions for obtaining different products with desired composition, biological activity and particle size from plant materials. The use of ultrasound to assist the production process steps is another topic that has been under investigation.

OBTAINING ANNATTO SEEDS PRODUCTS CONTAINING BIXIN: EXTRACTS

The recovery of bixin from annatto seeds has been traditionally performed by conventional solid-liquid extraction, which usually requires a long processing time that demands large amounts of organic solvents, alkali aqueous solutions or oils at higher temperatures. Nevertheless, these methods, in general, can promote the degradation of such compound [8, 11].

To overcome these drawbacks, conditions of short extraction times using pressurized solvent methods, such as supercritical fluid extraction (SFE) and pressurized liquid extraction (PLE) methods, have been successfully used to obtain pigment-rich extracts [13, 16]. SFE has important advantages over traditional extraction techniques. With the SFE method, low volumes of organic solvents, if any, are employed, and a solvent-free extract can be obtained with this method [17]. On the other hand, the use of PLE generally results in higher extraction yields in reduced time [18]. Moreover, high-pressure extraction has other advantages that should be considered. For example, native enzymes, which degrade some compounds, are inhibited by increasing extraction pressures and/or the addition of CO_2, the most common supercritical solvent [19].

The supercritical fluid state is characterized by fluid being above its critical temperature (Tc) and critical pressure (Pc). Manipulating the temperature and pressure of the fluid can solubilize the compound or class of compounds of interest and selectively extract it [20].

A major advantage of PLE over conventional solvent extraction methods conducted at atmospheric pressure is that pressurized solvents remain in a liquid state well above their boiling points, allowing for high-temperature extraction. Higher extraction temperature can improve analyte solubility in the solvent and improve its desorption from the raw material matrix [21].

The extraction method with pressurized fluids uses lower volumes of solvent to conventional extraction (Soxhlet, etc.) and is faster (extraction time varies between 10 and 20 minutes) [22]; thus, decreasing degradation of thermolabile compounds.

Although various methods for annatto seeds extract production have been investigated by different research groups, in this chapter our purpose is to discuss the results obtained and present futures perspectives related to the use of pressurized fluid-based techniques by our research group (LASEFI/DEA/FEA/UNICAMP). The same consideration was done in the sections related to the extract processing (fractionation, emulsification, encapsulation, etc.).

Annatto seeds (*Bixa orellana* L.) used in all experiments were cultivated in the state of São Paulo (variety Piave) and supplied by the Agronomical Institute of Campinas (IAC), Department of Agriculture and Supply of the State of São Paulo. Immediately after acquisition, the sample was identified and stored at 255 K, protected from light, until further analysis. All reagents and solvents used were of analytical grade.

By Supercritical Fluid Extraction (SFE)

Carbon dioxide (CO_2) is most solvent used for extraction, fractionation and encapsulation purposes due to the supercritical region can be achieved at moderate pressures and temperatures (Tc = 304.2 K, Pc = 7.38 MPa); therefore, working with supercritical CO_2 it is possible to carry out the process at near-ambient temperatures, avoiding the degradation of thermolabile substances [23-25].

Figure 1 shows a simplified flow diagram for a Supercritical Fluid Extraction (SFE) unit. The raw material (comminuted plant with moisture content bellow 20% if CO_2 is the solvent) is accommodated inside the extractor vessels forming a fixed bed. The supercritical fluid (SCF) flows through the fixed bed, the solute is carried by the SCF and the solvent/solute mixture is separated in the flash tank. Heating and cooling are required in various step of the process. In some cases, the solubility of compounds in supercritical CO_2 is not good. This can be overcome by the addition of cosolvent, usually a polar solvent such as water, ethanol, methanol, etc. to the supercritical solvent; this affects the properties of the fluid phase because of the strong interactions among the solute, the solvent, and the cosolvent [25].

SFE process has been applied for obtaining annatto seeds extracts by different research groups. Considering that the solubility of bixin in pure supercritical CO_2 is low, alternatives have been proposed as the use of organic solvents as cosolvents. But, using pure supercritical CO_2 can be advantageous for the extraction of several bioactives, providing greater selectively than the corresponding process with organic solvents [1, 26].

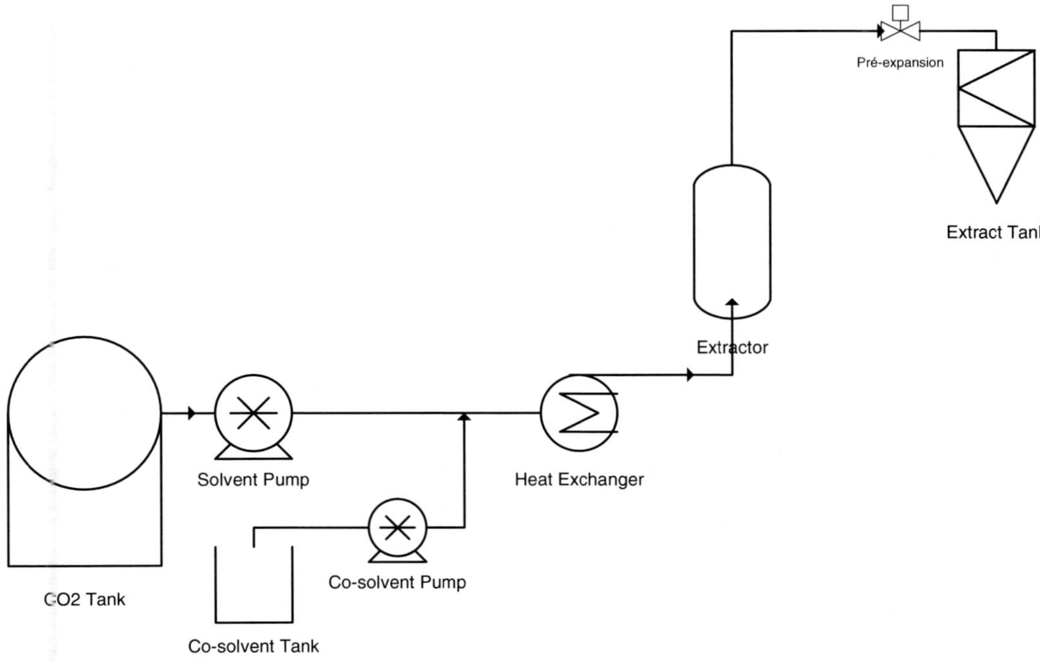

Figure 1. Simplified flow diagram for a Supercritical Fluid Extraction (SFE) unit.

Considering the context of low content of bixin in the SFE extracts; the relatively high lipid content (up to 4.8% have been reported in literature) of annatto seeds, which has substances (mainly tocotrienols) with functional properties, our research group has envisioned a two-step process for the extraction of both fractions: tocotrienols and bixin. Albuquerque and Meireles [26] demonstrated the feasibility of using pure supercritical CO_2 for obtaining the lipid-rich fraction of annatto seeds prior to the extraction of bixin with a more suitable extracting solvent. Employing similar ratio of solvent mass to feed mass (S/F) the extraction yield of the Soxhlet process using petroleum ether was (3.5 ± 0.1)% while the yield for the SFE process (333 K/40 MPa) was (3.58 ± 0.04)%; comparing these yields the efficiency of the SFE process is approximately 100%. It is interesting to observe that the total lipid content of annatto seeds was 3.7% a value close to that obtained for the SFE process.

In this study, the SFE process parameters were optimized. The highest yield of extract and bixin were obtained at 333 K and 40 MPa. Process conditions for low yields of bixin can be selected (333 K and 20 MPa) so as to obtain an amount of extract with the lowest level of pigment possible, enabling subsequent re-extraction of bixin from the seeds. At the end of the extraction it was observed that the bixin powder was easily detached from annatto seeds at the slightest friction, indicating that the defatting of annatto seeds using supercritical carbon dioxide extraction additionally may help the further re-extraction of bixin with a more suitable extracting solvent and method. Thus, further experiments were done in such direction.

By Pressurized Liquid Extraction (PLE)

By using conventional liquid solvents at elevated temperatures and pressures, Pressurized Liquid Extractions (PLE) increases the efficiency of the extraction process. Increased temperature accelerates the extraction kinetics, and elevated pressure keeps the solvent below its boiling point, thus enabling rapid extractions. Sometimes referred to as pressurized solvent extraction (PSE) and accelerated solvent extraction (ASE), PLE has been successfully used for the extraction of unstable compounds from various plant materials [22]. Figure 2 shows a simplified flow diagram for a PLE unit.

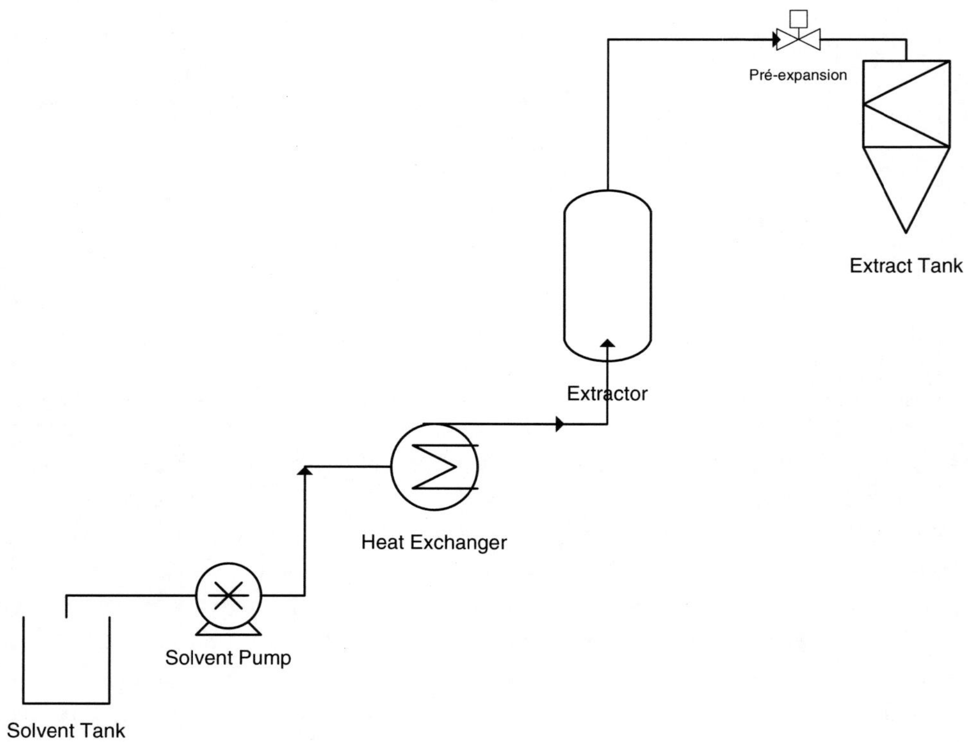

Figure 2. Simplified flow diagram for a Pressurized Liquid Extraction (PLE) unit.

In order to find a more suitable extracting solvent and method for bixin recovery from annatto seeds solvents with different polarities were evaluated employing the PLE method. PLE was chosen due to the easiness to be used in re-extraction of the annatto seeds defatted by SFE, avoiding plant material handling, transporting and storing. The co-solvent pump of the SFE unit could be used to pump the liquid solvent.

Recently, we have validated a pressurized solvent extraction system designed and constructed by our research group that can be independently used for supercritical fluid extraction (SFE) and pressurized liquid extraction (PLE) processes. Fractionated extractions of Jabuticaba skins were produced in two steps [27]. In the first step (PLE), pressurized ethanol was used to extract polar compounds including anthocyanin pigments. In the second step, pure supercritical CO_2 was used to recover low-polarity, CO_2-soluble compounds. One extract (PLE), presenting a purple color, was rich in anthocyanin pigments, and the other extract (SFE), presenting a yellow/green color, was rich in lipophilic compounds including essential oils and less polar flavonoid compounds.

Aiming similar approach but using PLE in the second step, different solvents water, ethanol, ethyl acetate, chloroform, dichloromethane, ethyl ether and acetone were evaluated employing the PLE method. The results indicate that moderately polar solvents, such as dichloromethane, ethyl acetate and ethanol, are better suited for the extraction of bixin from annatto seeds. These results are in accordance with previous reports where bixin, a lipophilic carotenoid, showed affinity for moderately polar solvents. An interest result obtained employing defatted annatto seeds by SFE was that the use of not fixed bed extraction method and water as extracting solvent produced much higher bixin yields that when fixed bed extraction methods, such as PLE, are employed [28]. Given that bixin solubility in water is very low it is indicated that the resulted product (bixin-rich extract in water) is a suspension obtained due to the observed easiness of recovery the bixin from the SFE processed seeds. The possible use of this process in the second step is also under evaluation taking into account that plant material handling, transporting and storing of the SFE seeds after the agitated bed extraction are necessary.

Thus, the proposed two-step process for the fractionation of lipophilic and bixin rich-extracts from annatto seeds would employ pure supercritical CO_2 in the first step and dichloromethane, ethyl acetate or ethanol under pressure in the next step. Since these organic solvents are highly miscible in supercritical CO_2, resulting in high volumetric expansion in supercritical CO_2, the use of processes that use supercritical CO_2 as anti-solvent, such as Supercritical Anti-solvent (SAS) fractionation, Supercritical Anti-Solvent (SAS) co-precipitation or Supercritical Fluid Extraction of Emulsion (SFEE) can be explored in a three-step process for annatto seeds processing with pressurized fluids in all steps. While the SAS fractionation process results in a more concentrated extract in the desirable compound or class of compounds, SAS co-precipitation produces coprecipitates containing the extract together with carrier materials and SFEE results in suspensions containing the extract. An interest point of these processes is that using dichloromethane, ethyl acetate or ethanol as solvent the remaining solvent in the products (concentrated extract, coprecipitate or suspension) is much lower than the concentration limit (around 10-50 ppm) than the concentration limit (600 ppm) [29].

By Supercritical Anti-solvent (SAS) Fractionation

Sometimes referred to as Supercritical Anti-solvent Fractionation (SAF), Supercritical Anti-solvent Extraction (SAE), Supercritical Anti-solvent (SAS) Precipitation, Supercritical Anti-Solvent (SAS) Fractionation, or simply SAS fractionation has been successfully used for obtaining more concentrated products. The pre-conditions to apply SAS fractionation are similar to the ones characteristics of supercritical antisolvent (SAS) micronization: the liquid solvent has to be very soluble in supercritical CO_2, whereas, the desirable compounds or class of compounds have to be completely not soluble in the SCF. The scope of the process is not the micronization, but the selective extraction of part of the solid compounds. These conditions can be frequently obtained since many organic solvents are readily soluble in supercritical CO_2 even at mild operating conditions and many high molecular weight solids show negligible solubilities in supercritical CO_2 especially at low CO_2 densities [30].

For our purpose, to obtain a more rich extract in bixin from the PLE solution (bixin-rich extract in dichloromethane, ethyl acetate or ethanol; ethanol will be preferred in such study due to be a GRAS - "Generally Recognized As Safe" solvent), a concentrated bixin fraction as the primary product, and an oil/organic solvent fraction as a secondary product would be obtained. Experiments will be done in order to achieve the more selective extraction of the undesirable part of PLE extract solution, concentrating the precipitated extract enriched in bixin. Figure 3 shows a simplified flow diagram for a SAS fractionation unit.

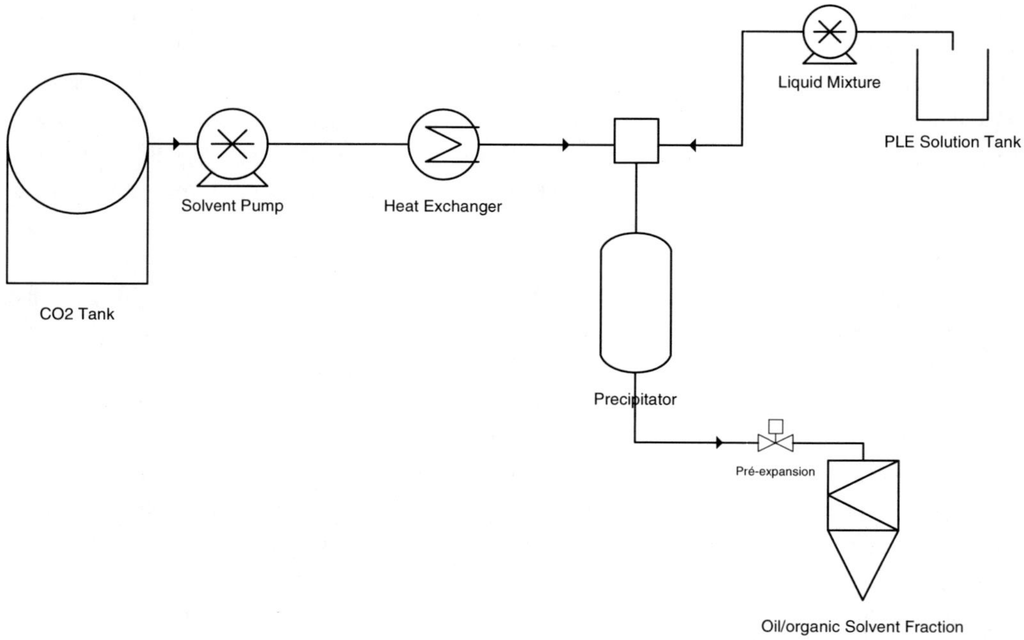

Figure 3. Simplified flow diagram for a Supercritical Anti-Solvent (SAS) fractionation unit.

Obtaining Annatto Seeds Products Containing Bixin: Coprecipitates

Another possibility for the third step is the use of Supercritical Anti-Solvent (SAS) co-precipitation process to obtain coprecipitates containing the bixin-rich extract from PLE together with carrier materials. Similarly to the SAS fractionation process, the PLE solution (bixin-rich extract in dichloromethane, ethyl acetate or ethanol) is put in contact with the supercritical CO_2 in order to promote the organic solvent removal and consequently solid compounds precipitation. The difference between both processes is the addition a carrier material into the PLE solution. Therefore, the organic solvent employed must solubilize the extract and also the carrier material used dichloromethane was used in the experiments due to the high solubility of PEG 10000 in this solvent.

The effect of mass ratio between bixin-rich extract and PEG on the co-precipitation process was evaluated by Santos and Meireles [31]. According to our findings and corroborating the literature data, a decrease in the mass ratio between core material and carrier material, i.e., an increase in the polymer concentration, led to the production of less agglomerated particles. Then, the smaller (1:10) mass ratio between bixin-rich extract and PEG was chosen. For the mass ratio of 1:2, it was visually observed that, probably, the amount of polymer was not sufficient to effectively encapsulate the amount of bixin-rich extract.

In other study by our research group, Santos et al. [32] demonstrated that bixin stability in the PLE extracts is improved by its coprecipitation with a carrier material. Also in this study the possibility of coupling PLE method to SAS co-precipitation, in such a way that PLE and SAS co-precipitation could be performed in only one step was demonstrated. Interestingly some differences were observed between the co-precipitate products obtained by this one step (on-line) process when compared with the usual two step process (when the PLE extract is not put in contact with the supercritical CO_2 directly). While the on-line process produced a darker capsule with deep red color, the capsule produced by the off-line process had a lighter orange coloration, indicating the degradation of bixin. The difference in coloration can be explained by the fact that bixin remains in an inert atmosphere throughout the extraction and co-precipitation in the on-line process. In the off-line process, during collection of the extract and after the extraction, bixin is exposed to light and air and therefore may be subjected to degradation. Furthermore, the results showed a faster degradation of bixin from capsule obtained by the off-line process (lower $t_{1/2}$) in all studied conditions. Experiments will be done in order to achieve by SAS co-precipitation besides bixin protection and controlled particle size a selective fractionation for obtaining a coprecipitated extract more enriched in bixin, adding additional value to this product. Figure 4 shows a simplified flow diagram for the PLE-SAS co-precipitation developed system.

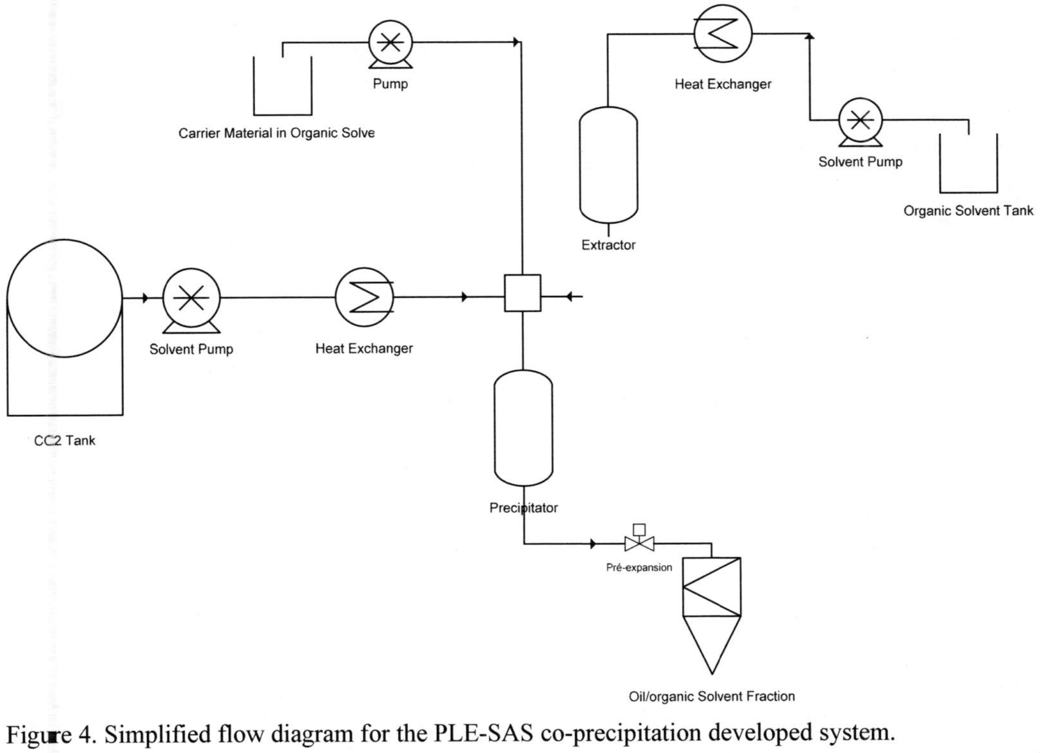

Figure 4. Simplified flow diagram for the PLE-SAS co-precipitation developed system.

Obtaining Annatto Seeds Products Containing Bixin: Nanosuspensions

The use of Supercritical Fluid Extraction of Emulsions (SFEE) to produce micro or nanosuspensions containing the bixin-rich extract from PLE also can be a option for adding value to the PLE extract. As the other two anti-solvent processes, SAS fractionation and SAS co-precipitation, this process has the same configuration been all of them performed in the same experimental apparatuses. The only difference is that an emulsion is pumped for organic solvent removal from the emulsion droplet instead of a solution containing extract with or without a carrier material. Since an emulsion need to be formed surfactant and an additional step is necessary [33]. In such a way our research group is studying different emulsification processes: 1) Emulsification by mechanical agitation; 2) Emulsification by ultrasound in batch or continuous manner combined or not under pressures and temperatures higher than atmospherics ones using different surfactants: 1) n-octenyl succinic anhydride (OSA)-modified starches; 2) Saponin-rich extracts from *Pfaffia glomerata* roots; 3) Tween-type compounds.

Recently we demonstrated using a different raw material the possibility of coupling PLE method to SFEE [34], which might be easily applied for the production in only one step of nanosuspensions containing the bixin-rich extract directly from PLE. The limitation of the PLE-SFEE process is related to the fact that only water-immiscible organic solvents can be used as extracting solvent. In this context, dichloromethane and ethyl acetate should be used to obtain the PLE bixin-rich extract solution that will be able to be emulsified with the aqueous surfactant solution. The potentially of the use of ultrasound irradiation to improve this continuous emulsification is under evaluation, indicating very promising perspectives.

The possibility of coupling to the PLE-SFEE the PLE process of aqueous saponin-rich extracts from *Pfaffia glomerata* roots is also under study by our research group. In this context, after annatto seeds defatting by SFE, PLE using dichloromethane or ethyl acetate as extracting solvent is performed for the remained solid material while water under high pressure and temperature flows through another extractor containing *Pfaffia glomerata* roots. At the end both flows are put in contact in a vessel with a ultrasound probe to promote emulsification of the organic phase in water with the aid of the surfactant compounds present in the *Pfaffia glomerata* roots. Then, the formed emulsion is put in contact with supercritical CO_2 for the removal of the organic solvent and consequently suspension production. Experiments can be done in order to achieve by SFEE besides bixin protection, controlled emulsion droplet size and suspension particle size a selective fractionation for obtaining a suspension containing an extract more enriched in bixin, adding additional value to this product. Figure 5 shows a simplified flow diagram for this hypothetical process.

Due to the similarities already presented PLE-SAS fractionation, PLE-SAS co-precipitation and PLE-SFEE processes were grouped end called by Organic solvent Extraction and Particle formation On-line (OEPO) due to the similarities to the WEPO (Water Extraction and Particle formation On-line) process developed by Ibanez group in Spain. They have developed a hyphenated process to obtain dried powders of extracts from natural sources also in one step. Similarly to ours WEPO also use PLE technique for bioactive compounds recovery, but employ water as extracting solvent. Since, in their case supercritical CO_2 are not suitable for solvent elimination due to the low solubility of water in CO_2, this fluid is used as a dispersion medium and hot N2 is used as drying agent. Santos et al. [34] demonstrated the versatility of our hyphenated-based process that uses pressurized fluids in both unit operations (extraction and particle formation) using *Pfaffia glomerata* roots as plant material.

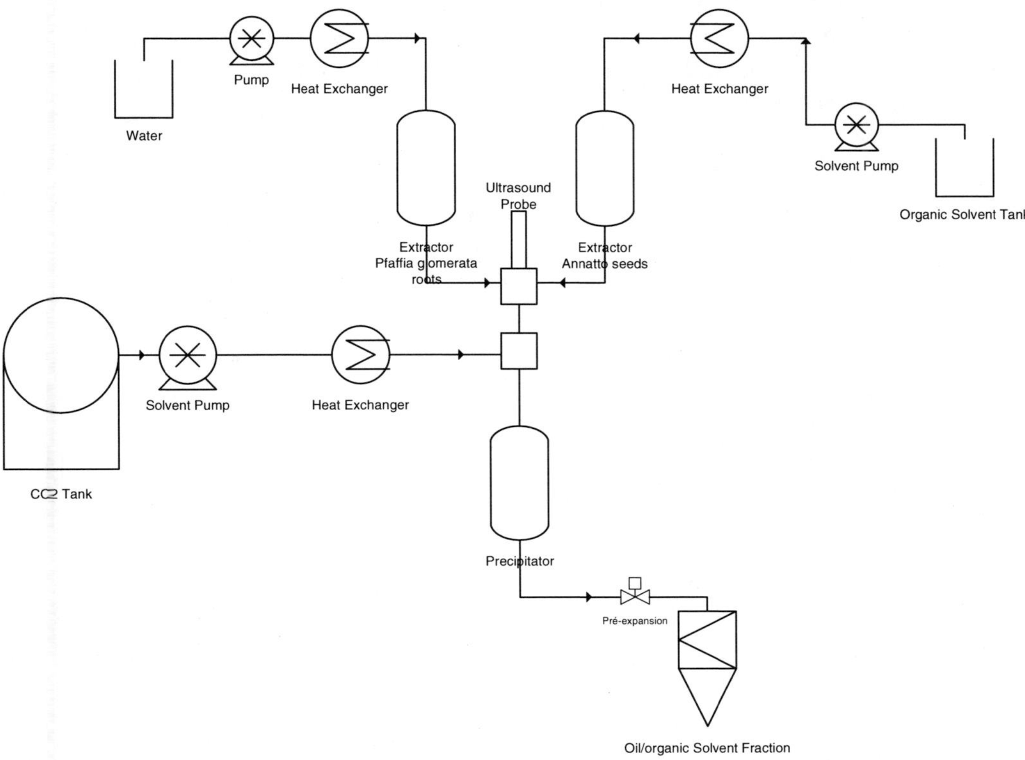

Figure 5. Simplified flow diagram for the hypothetical process based on the coupling to the PLE-SFEE the FLE process of aqueous saponin-rich extracts from *Pfaffia glomerata* roots.

OBTAINING ANNATTO SEEDS PRODUCTS CONTAINING BIXIN: NANOEMULSIONS

With a focus on increase even more the high added value of the SFE extract obtained using pure supercritical CO_2 different approaches are under evaluation by our research group. An interesting option should be a formulation that allows protecting the bioactive compounds (mainly tocotrienols) from high temperatures, oxidation and UV light. In the food, pharmaceutical, agricultural, and cosmetic industries, there is strong interest in creating new products based on oily systems in which emulsions can be used as delivery systems. Thus, we are planning optimize the emulsification process to achieve emulsion droplet size in the range of nanometers.

Studying the emulsification process of glove love essential oil obtained by supercritical fluid extraction (SFE) Santos et al. [35] demonstrated that emulsification by ultrasonication employing similar energy density are comparable in terms of emulsion droplet size and stability with those of common mechanical emulsifying techniques. In another work we also demonstrated the possibility to perform an emulsification with the PLE aqueous extract flow from a saponin source, such as *Pfaffia glomerata* roots. Named by our research group as Emulsion from Pressurized Liquid Extraction (EPLE), this process might be coupled to the SFE process for obtaining lipophilic fraction from annatto seeds [36]. This process would provide the continuous emulsification of this SFE lipophilic fraction with the PLE aqueous extract flow with the aid of ultrasound irradiation. A remarkable point of this process is that the emulsification step will occur under a gas-expanded liquid (annatto seeds oil and water – liquid state; CO_2 supercritical state) atmosphere. Figure 6 shows a simplified flow diagram for this hypothetical process.

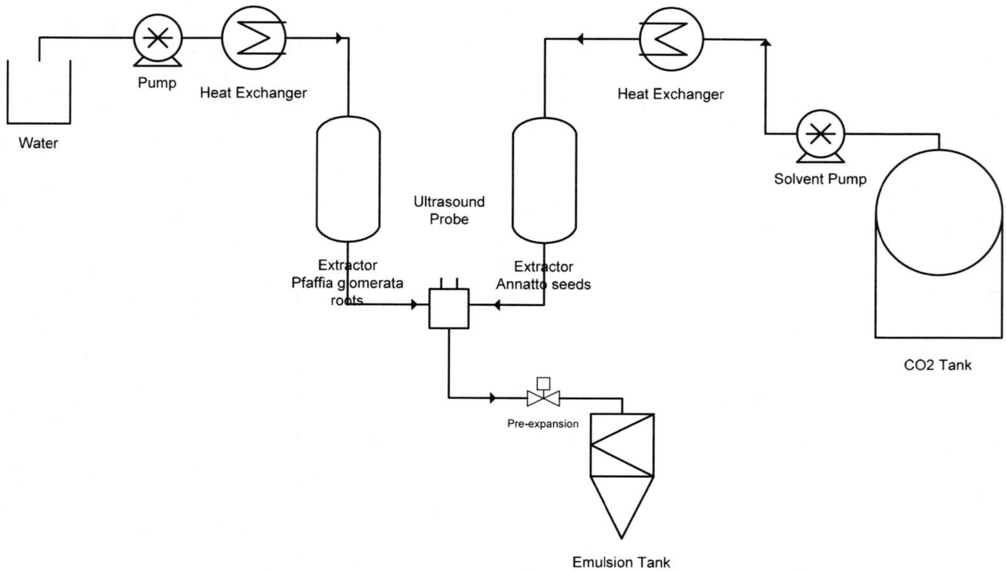

Figure 6. Simplified flow diagram for the hypothetical process based on the coupling to the Emulsion from Pressurized Liquid Extraction (EPLE) process to the SFE process for obtaining lipophilic fraction from annatto seeds.

CONCLUSION

Recent works done and under evaluation by our research group (LASEFI/DEA/FEA/UNICAMP) for obtaining annatto seeds processing with pressurized fluids were presented. Different strategies for obtaining annatto seeds products: extracts, coprecipitates, nanosuspensions and nanoemulsions aiming higher yields, low manufactured cost with concomitant minimal carotenoid degradation are discussed. Fractionated extractions using different pressurized fluids in each step linked with the coupling of extraction and particle formation in one single process and the use of ultrasound irradiation to assist the emulsification process step indicated a great potential to be explored for adding value to the annatto seeds and other plant materials.

ACKNOWLEDGMENTS

Diego T. Santos is thankful to FAPESP (processes 10/16485-5; 12/19304-7) for the postdoctoral fellowships. M. Thereza M. S. Gomes, Liara M. Rodrigues and R. Abel C. Torres would like to thank CNPq for the fellowships. The authors acknowledge the financial support from CNPq and FAPESP (processes 09/17234-9; 12/10685-8).

REFERENCES

[1] Chao, R. R., Mulvaney, S. J., Sanson, D. R., Hsieh, F., Tempesta, M. S. (1991). Supercritical CO_2 Extraction of Annatto (*Bixa orellana*) Pigments and Some Characteristics of the Color Extracts. *Journal Food Science*, 56, 80-83.

[2] Food and Agriculture Organization of the United Nations (FAO) (2006). Annatto extracts: chemical and technical assessment, in: 67th Joint FAO/WHO Expert Committee on Food Additives, Rome, Italy. Available in: ftp://ftp.fao.org/ag/agn/jecfa/ctaannatto.pdf.

[3] Souza, E. C. (1998). Chemical and Chemico-physical study of Pigments of Urucum (*Bixa orellana* L.) using a Simplified Methodology of Extraction, M.Sc.Thesis, Universidade Federal de Lavras, Lavras, Brazil.

[4] Preston, H. D., Rickard, M. D. (1980). Extraction and chemistry of annatto. *Food Chemistry*, 5, 47-56.

[5] Montenegro, M. A., Rios, A. O., Nazareno, M. A., Mercadante, A. Z., Borsarelli, C. D. (2004). Model studies on the photosensitized isomerization of bixin. *Journal of Agricultural and Food Chemistry*, 52, 367-373.

[6] Cooper, D. A. (2004). Carotenoids in health and disease: Recent scientific evaluations, research recommendations and the consumer. *Journal of Nutrition*, 134, 221S-224S.

[7] Rodriguez-Amaya, D. B., Kimura, M., Godoy, H. T., Amaya-Farfan, J. (2008). Updated Brazilian database on food carotenoids: Factors affecting carotenoid composition. *Journal of Food Composition and Analysis*, 21, 445-463.

[8] Scotter, M. J. (1995). Characterization of the Coloured Thermal Degradation Products of Bixin from Annatto and a Revised Mechanism for their Formation. *Food Chemistry*, 53, 177-185.

[9] Scotter, M. J., Wilson L. A., Appleton, G. P., Castle, L. (2000). Analysis of Annatto (*Bixa orellana*) Food Coloring Formulations. 2. Determination of Aromatic Hydrocarbon Thermal Degradation Products by Gas Chromatography. *Journal of Agricultural and Food Chemistry*, 48, 484-488.

[10] Gloria, M. B. A., Vale, S. R., Bobbio, P. A. (1995). Effect of water activity on the stability of bixin in an annatto extract-microcrystalline cellulose model system. *Food Chemistry*, 52, 389-391.

[11] Zechmeister, L., Escue, R. B. (1994). A stereochemical study of methylbixin. *Journal American of the Chemical Society*, 66, 322-330.

[12] Montenegro, M. A., Rios, A. O., Mercadante, A. Z., Nazareno, M. A., Borsarelli, C. D. (2004). Model Studies on the Photosensitized Isomerization of Bixin. *Journal of Agricultural and Food Chemistry*, 52, 367-373.

[13] Silva, G. F., Gamarra, F. M. C., Oliveira, A. L., Cabral F. A. (2008). Extraction of bixin from annatto seeds using supercritical carbon dioxide. *Brazilian Journal of Chemical Engineering*, 25, 419-426.

[14] Shahidi, F., Han ,X. Q. (1993). Encapsulation of food ingredients. *Critical Reviews in Food Technology*, 33,501-504.

[15] Madene, A., Jacquot, M., Scher, J., Desobry, S. (2006). Flavour encapsulation and controlled release – a review, *International Journal of Food Science*, Technology, 41, 1-21.

[16] Arapitsas, P., Turner, C. (2008). Pressurized solvent extraction and monolithic column-HPLC/DAD analysis of anthocyanins in red cabbage. *Talanta*, 74, 1218-1223.

[17] Santos, D. T, Meireles, M. A. A. (2011). Extraction of Volatile Oils by Supercritical Fluid Extraction. *Recent Patents on Engineering*, 5, 17-22.

[18] Ju, Z. Y., Howard, L. R. (2003). Effects of solvent and temperature on pressurized liquid extraction of anthocyanins and total phenolics from dried red grape skin. *Journal of Agricultural and Food Chemistry*, 51, 5207-5213.

[19] Seabra, I. J., Braga, M. E. M., Batista, M. T., Sousa, H. C. (2010). Effect of solvent (CO_2/ethanol/H_2O) on the fractionated enhanced solvent extraction of anthocyanins from elder berry pomace. *Journal of Supercritical Fluids*, 54, 145-152.

[20] Watters: the science of what's possible. Supercritical Fluid Extraction (SFE) Systems (2013). Available in http://www.waters.com/waters/nav.htm?cid=10146521& locale=pt_BR.

[21] Richter, B. E., Jones, B. A., Ezzell, J. L., Porter, N. L. (1997). Accelerated Solvent Extraction: A Technique for Sample Preparation. *Analytica Chemistry*, 68, 1033–1039.

[22] Santos, D. T., Veggi, P. C., Meireles, M. A. A. (2012). Optimization and economic evaluation of pressurized liquid extraction of phenolic compounds from jabuticaba skins. *Journal of Food Engineering*, 108, 444-452.

[23] Santos, D. T., Meireles, M. A. A. (2010). Carotenoid Pigments Encapsulation: Fundamentals, Techniques and Recent Trends. *The Open Chemical Engineering Journal*, 4, 42-50.

[24] Cocero, M. J., Martín, A., Mattea, F., Varona, S. (2009). Encapsulation and co-precipitation processes with supercritical fluids: fundamentals and applications. *Journal of Supercritical Fluids*, 47, 546-555.

[25] Pereira, C. G., Meireles, M. A. A. (2010). Supercritical Fluid Extraction of Bioactive Compounds: Fundamentals, Applications and Economic Perspectives. *Food and Bioprocess Technology*, 3, 340-372.

[26] Albuquerque, C. L.C., Meireles, M. A. A. (2012). Defatting of annatto seeds using supercritical carbon dioxide as a pretreatment for the production of bixin: Experimental, modeling and economic evaluation of the process. *Journal of Supercritical Fluids*, 66, 86-95.

[27] Santos, D. T., Albuquerque, C. L.C., Meireles, M. A. A. (2011). Antioxidant dye and pigment extraction using a homemade pressurized solvent extraction system. *Procedia Food Scienc*e, 1, 1581-1588.

[28] Rodrigues, L. M., Alcázar-Alay, S. C., Gomes, M. T. M. S., Meireles, M. A. A. (2013). Pressurized liquid extraction (ple) of bixin from defatted annatto seeds. III Iberoamerican Conference on Supercritical Fluids (PROSCIBA), *PROSCIBA 2013 proceedings*

[29] Gomes, M. T. M. S., Santos, D. T., Meireles, M. A. A. (2012). Trends in Particle Formation of Bioactive Compounds Using Supercritical Fluids and Nanoemulsions. *Food and Public Health*, 2, 142-152.

[30] E. Reverchon, I. Marco. (2006). Supercritical fluid extraction and fractionation of natural matter. *Journal of Supercritical Fluids*, 38, 146-166.

[31] Santos, D. T., Meireles, M. A. A. (2013). Micronization and encapsulation of functional pigments using supercritical carbon dioxide. *Journal of Food Process Engineering*, 36, 36-49.

[32] Santos, D. T., Gomes, M. T. M. S., Vardanega R., Rostagno, M. A., Meireles, M. A. A. (2013). Integration of Pressurized Fluid-based Technologies for Natural Product Processing. In: Rostagno, M. A. and Prado, J. (Eds.), Natural Product Extraction: Principles and Applications. Royal Society of Chemistry.

[33] Santos, D. T., Martín, A., Meireles, M. A. A., Cocero, M. J. (2012). Production of stabilized sub-micrometric particles of carotenoids using supercritical fluid extraction of emulsions. *Journal of Supercritical Fluids*, 61, 167-174.

[34] Santos, D. T., Barbosa, D. F., Broccolo, K.. Gomes M. T. M. S, Vardanega, R., Meireles, M. A. A. (2012). Pressurized Organic Solvent Extraction with On-line Particle Formation by Supercritical Anti Solvent Processes. *Food and Public Health*, 2, 231-240.

[35] Santos, D. T., Barbosa, D. F., Vardanega, R., Albarelli, J. Q. Meireles, M. A. A. (2013). Experimental and Simulation Study on Formulation of Clove Essential Oil Products Using Alternative Surfactant. *Journal of Colloid Science and Biotechnology*, in press.

[36] Santos, D. T., Barbosa, D. F., Vardanega, R., Gomes, M. T. M. S., Meireles, M. A. A. (2013). Novel method to produce emulsions containing essential oils from saponin-rich pressurized aqueous plant extracts. *Journal of Colloid Science and Biotechnology*, in press.

In: Processed Foods
Editors: Chloe M. Gagne and Daniel B. Jones

ISBN: 978-1-62808-688-1

Chapter 9

PRINCIPLES OF QUALITY ANALYSIS AND CRITICAL CONTROL POINTS APPLICABLE TO FOOD PROCESSING

Wenbiao Wu[1,*], Yixing Yang[2] and Wei Huang[3]

[1]College of Food Science, Southwest University,
Tian Sheng Qiao, Beibei, Chongqing, PRC
[2]School of Public Health, Dali University, Dali, PRC
[3]School of Forests and Bioscience, Chongqing University of Arts and Sciences,
Honghe Road, Yong Chuan, Chongqing, PRC

ABSTRACT

Consumer's choice of a food product is usually based on its good quality. Food manufacturers always produce a food product that is demanded by consumers. The target of QACCP system is to assure the processing of the food product with its quality as high as possible and its quality controlling cost as low as possible. This would attract food manufacturers. Implementation of QACCP follows eight principles including *Principle 1-Establishment of the principles for quality analysis*, *Principle 2-Determination of critical factors according to food standard established*, *Principle 3-Recognition of critical control points by quality analysis*, *Principle 4-Establishment of critical limits for each critical control point*, *Principle 5-Creation of measures suitable for monitoring the status of critical limit for each critical control point*, *Principle 6-Creation of procedures for corrective action*, *Principle 7-Creation of procedures for verifying the QACCP system* and *Principle 8-Creation of procedures for keeping record or memorandum document.* By strictly following the 8 principles of QACCP described above, it is reasonable to believe that any food product processed can at least meet its standard previously established. It is believed that the tables provided are well designed, quite accurate and easily operative. They should be very helpful for putting the 8 principles of QACCP into practice. Therefore, the application of QACCP to food manufacturing may be very

---

* Correspondent author. Adress: College of Food Science, Southwest University, 216 Tian Sheng Qiao, Beibei, 400716 Chongqing, PRC. Tel: 0086 23 68259815, E-mail address: wbwu62@yahoo.com.

perspective. The 8 principles of QACCP may also be helpful for food scientists and engineers. They may be included in a text book to teach students in food quality control.

Keywords: Principles, QACCP system, Quality control, Food product

1. INTRODUCTION

Manufacturers produce food products mainly because consumers demand them. The principle purpose of buying a food product whose dietary safety has already been proven by scientific research and accepted by consumers is to get enough nutrients essential to the life activities of their bodies. Generally, on the premise of nutritional requirements and dietary safety already met, consumer's choices could be made according to the sensory properties, such as taste, color, smell, texture, form and morpha of a food product including the nature of package. Other factors, such as that affects the expenditure of buying a food product (e.g. its packing quantity) should also always be cared about by consumers. Therefore, the assemblage of properties or characteristics of a food product, which determines the degree of acceptability of the food product to consumers, may be defined as Food Quality (Kramer and Twigg, 1968).

It is legally mandatory that any aspect of its quality relating to its properties and packing specifications, etc, is assigned by standards before the production of any food (see *Chapter 7 Basic guidelines for establishing food standards*). A food standard may be established by the federal, state or municipal agencies, companies, etc. However, no matter who has established the food standard, how to meet it during processing is always a problem that needs to be tackled. Therefore, a proper quality control or assurance procedure is necessary for producing a food product being conformable to its standards. The quality control or assurance procedure should cover all stages of the food manufacturing, purchasing of raw material, storing, transporting and selling. During these stages, quality analysis is also essential. According to its variety, one, or several, or all the properties or characteristics that include nutrients, hazards, sensory properties and package of a food product should be strictly analyzed and controlled during its processing. Nutrients include amino acids, proteins, saccharides, fatty acids, oils or fats, essential nutritional minerals and vitamins. Hazards include toxic inorganic and organic substances, pathogenic micro-organisms. Sensory properties include taste, color, smell, texture, form and morpha. Package quality is related to quantity, size, materials, strength, etc.

From past literatures, although it can be seen that Hazards Analysis and Critical Control Point (HACCP) is widely employed as a systematic preventive approach to food safety that addresses physical, chemical, and biological hazards as a means of prevention rather than finished product inspection (FAO/WHO, 2007; United States Food and Drug Administration, 2007; Mayes and Mortimore, 2001; Bauman, 1995). This is only one aspect of whole food quality control and formal prerequisite programs should be used to support the implementation of HACCP in food processing (Wallace and William, 2001). Peariso (2006) promoted Quality Assurance of Critical Control Points (QACCP) plan for preventing foreign material contamination of foods. Alli (2004) introduced the principles and practices of food industry good manufacturing practices (GMPs), HACCP, and the ISO 9000 quality management system standards. The same author also stated that the field of food quality assurance has evolved substantially over the past decade, and certain key developments have become widely accepted, which include Quality Systems (e.g., ISO 9000) and HACCP. Vasconcellos (2004) described the basic concepts and principles behind Total Quality Management and examples of Quality Assurance programs that can be applied to the food industry.

Furthermore, the long-term solution for food companies to sustain an international demand for their products depends on building up the trust and confidence of consumers in the quality of their food supply systems. This requires improvements within national food control systems and within the industry's food quality programs (FAO, 1999). 'Farm to table' process control to manage both quality and safety is increasingly in demand in developed countries, and new institutions are evolving to certify production practices (Unnevehr and Jensen, 1999). However, the direct monitoring of 'Farm to table' processes may be prohibitively expensive in most cases. For the full control of the quality of any food, it is also necessary to determine whether quality problems are due to lack of care and effort by the primary producer or to factors outside his control (Otsuka, et al., 1992; Prendergast, 1999).

Those reports mentioned above imply that the establishment of a proper and efficient food quality control or assurance procedure as well as quality analysis is still necessary. Quality analysis and critical control points (QACCP) applicable to food processing is designed to solve the problem described above. Quality analysis is essential not only for setting out the accurate critical limits that should be included in the standard for any property and packing specification of a food product, but also for meeting the standard during processing. Establishment of critical control points is to lower the cost of controlling the quality of the food product quality to meet its standard as possible as ever it can.

Therefore, the objective of this article is to introduce the principles of QACCP, which have never been reported in literatures. QACCP system which takes into the consideration of nutrients, hazards, sensory properties and package of the food product together should be very efficient for assuring the quality of the food product during processing, transporting, storing and marketing.

2. Principles of QACCP

The principal aim of creating a QACCP system is to reduce the food quality assuring cost to a reasonable level and simplify its control procedures as possible as ever they can. The creation of any particular QACCP for a particular food product should follow the same principles. These principles are designed to guide the establishment of implementing procedures for whole quality control of the food product and their operations during manufacturing, transporting, storing and marketing. The QACCP created by following these principles should be able to assure that the food product being processed and sold is equal to or above its standards previously established.

It should be noted that there should be established standards for the food product that is going to be manufactured and sold before drafting a particular QACCP. The food standards should be assigned by an appropriate authority, such as the federal, state or municipal agencies, or companies. If the food product that is going to be manufactured and sold has newly been developed and its standard has not been established, the first work that has to be carried out is to establish its standard before drafting a particular QACCP.

The principle –"*Creation of procedures for keeping record or memorandum document"* of QACCP has been developed on the basis of that of HACCP (Mortimore and Emond, 2008). Other principles are established on the basis of analyzing the characteristics of food manufacturing technology and related scientific research outputs.

2.1. Establishment of the Principles for Quality Analysis

Any item of nutrients, hazards, sensory properties, packaging parameters, etc, which has been assigned in the standard of a particular food product being processed (See *Chapter 7 Basic guidelines for establishing food standards*), should be accurately analyzed and monitored during its processing. Quality analysis procedures (including sampling and analytical methods) may also be assigned in the standard. These assigned standards should also be strictly followed. In general, it should be noted that quality analysis procedures employed in QACCP system always follow the timely updated standard.

If the employment of a quality analysis method that is not assigned in the standard in QACCP system is inevitable, at least it should be one that has been published in authoritative journals, books or documents, or official standard methods bulletins ratified by the federal, state or municipal agencies or authoritative organizations, etc. Its suitability being applied to QACCP system should first be well evaluated and then it is included in the updated standard before putting it into practice. Of cause, research and development on new analytical methods, which are preparing for future should be friendly welcomed. However, the application of any newly developed method to QACCP system should always follow the principle described above.

Managing staff and operating technicians working on online monitoring or laboratory analysis of the quality of the food being processed should be well disciplined. Especially, they should be familiar with the quality analysis procedures or methods employed in QACCP system both theoretically and operatively.

According to the characteristics of the food product and its processing procedures, proper online monitoring and laboratory analysis may be implemented simultaneously, or just one of them is employed solely. Proper online monitoring site and method or laboratory procedures should be determined on the condition of that the quality of the food product being processed can be ensured to at least meet the established standard.

2.2. Determination of Critical Factors According to Food Standard Established

The food product should contain certain amount of nutrients with safe level of hazard substances and have acceptable sensory properties. It should also be properly packed, stored and transported. These are normally appointed by the standard of the food product previously established. Any item in relation to nutrient, sensory property, hazard substance, or packaging parameter, which has been assigned in the standard, is the critical factor for the good quality control of the food product being processed or sold.

Table 1. Analyzing table of critical factor (CF) relating to the quality of the food product being processed, transported, stored and sold[a]

Factors	Is it significant?	Reasons for being significant[b]	Is it CF?	Reasons for being CF[c]
Nutrients				
e.g. Protein content	Yes or No		Yes or No	
e.g. Lipid content	Yes or No		Yes or No	
e.g. Saccharide content	Yes or No		Yes or No	
e.g. Fe content	Yes or No		Yes or No	
e.g. Vitamin A content	Yes or No		Yes or No	
..	Yes or No		Yes or No	
Hazards				
e.g. Pb content	Yes or No		Yes or No	
…	Yes or No		Yes or No	
Sensory properties				
e.g. Color	Yes or No		Yes or No	
…	Yes or No		Yes or No	
Package				
e.g. Quantity	Yes or No		Yes or No	
…	Yes or No		Yes or No	

[a] Any food quality control method must follow food standard that is mandatory. Different foods have different standards. Therefore, the factors that should be included in the analyzing table of critical factor related to a particular food must conform to its standard established.

[b] The factors related to quality in a particular food such as Vitamins may be very susceptible to processing or other conditions. The more susceptible they are, the more significant they are.

[c] If the factors related to quality in a particular food have been appointed by its standard, they must be critical factor.

Table 1 can be used for the purpose of establishing the critical factor in practice. The critical factor must be strictly controlled during manufacturing, transporting, storing and marketing in order to assure that the food product being processed or sold is equal to or above the standards previously established. For example, protein content is the only critical nutritional factor for a vegetable protein product (protein content > 40%) or soy protein isolate (protein content > 90%) while other nutrients such as vitamins or minerals are not strictly required (FAO/WHO, 1989a, 1989b). The concrete cases of determining the critical hazard, sensory property, or package factor are quite similar to the critical nutritional factor described above.

2.3. Recognition of Critical Control Points by Quality Analysis

Many points or steps or procedures may affect the critical factor of the quality of the food product during manufacturing, storing, transporting and selling. These may include purchase of the raw materials, processing procedures, factory managing measures, transportation or storage conditions, etc. Therefore, this step determines the Critical Control Point (CCP) of the critical factors for making the food product with good quality. The CCP can be a point, step, or procedure in a food manufacturing, transporting, storing or marketing process at which control can be applied and, as a result, the critical factors including nutrient content, hazard substance content, sensory properties, or packaging quality that determine the quality of the food product can be assured to the acceptable levels of the standards previously established. Table 2 can be used for this purpose.

Table 2. Analyzing table of critical points (CCP) relating to the quality of the food product being processed, transported, stored and sold[a]

Control Points[b]	CF[c] being affected	Is it significant?	Reasons for being significant[d]	Is it CCP?	Reasons for being CCP[e]
Raw material					
e.g. Variety	e.g. Protein content	Yes or No		Yes or No	
…	…	Yes or No		Yes or No	
Processing					
e.g. pH	e.g. Protein content	Yes or No		Yes or No	
…	…	Yes or No		Yes or No	
Transporting					
e.g. Temperature	e.g. Microoganism	Yes or No		Yes or No	
…	…	Yes or No			
Storing					
e.g. Time	e.g. V_C content	Yes or No		Yes or No	
…	…	Yes or No			
Marketing					
e.g. Time	e.g. V_C content	Yes or No		Yes or No	
…	…	Yes or No		Yes or No	

[a] After the critical factors of the quality of a food product have been determined in Table 1, the operating point that can result in the variation of these critical factors during processing, transporting, storing and marketing can be determined and filled in this table.

[b] The control point can be any point on the production including the purchase of raw materials, transporting, storing and marketing flow-line of a product. These points can be determined according to knowledge, experience, laboratory-or plant-scale experiments, etc.

[c] These critical factors can be changed by the variation of a relating control point. They can be determined according to knowledge, experience, laboratory-or plant-scale experiments, etc.

[d] The more effective on CF levels a control point is, the more significant it is.

[e] If control is applied at a point as a result, the critical factors including nutrient content, hazard substance content, sensory properties, or packaging quality that determine the quality of the food product can be assured to the acceptable levels of the standards previously established, this point is CCP.

2.4. Establishment of Critical Limits for Each Critical Control Point

The critical limit is the maximum or minimum value of the processing parameters, or transporting, storing and marketing conditions, etc to which the critical factors must be controlled at the critical control point to assure that the critical quality value of the food product is equal to or better than the acceptable levels appointed by the standards. The processing or operational or manufacturing parameters, raw material or managing or transporting or storing or selling conditions may include temperature, time, water activity, humidity, pH, concentration, recipe, etc (including chemical or physical parameters). The critical limit can usually be determined by laboratory- or pilot-plant-scale experiments.

2.5. Creation of Measures Suitable for Monitoring the Status of Critical Limit for Each Critical Control Point

Variation of the value of the processing parameters, or transporting, storing and marketing conditions, etc at critical control point definitely results in changes in the critical factor and consequently the alteration of the quality of the food product. This must be monitored to ensure that the process is under control at each critical control point. For achieving this, it is necessary for a quality administrator to establish well planned measures monitoring the status of critical limit for each critical control point in an appropriate sequence. The measures may include the appointment of efficient chemical or physical determination methods, monitoring procedures and personnel management. Continuous online monitoring is the best. If this is not possible, adequate intermittent frequency of monitoring should be employed. Monitoring results must be recorded in documents relating to CCP monitoring and signed by monitor, administrator or at least ≥2 persons related. It is required that each monitoring procedure and its frequency be listed in the QACCP plan (Table 3).

2.6. Creation of Procedures for Corrective Action

If monitoring indicates a deviation from an established critical limit for a critical control point, corrective actions must be taken immediately. For achieving this, it is necessary for a quality administrator to establish well planned measures correcting the deviated critical limit for this critical control point.

The measures may include the appointment of efficient corrective methods, or procedures and personnel management. The identified corrective actions should be verified practicably efficient and written in a plant's QACCP plan. Corrective actions are intended to ensure no injurious to the quality of the food being processed, transported and sold. Otherwise the food product with adulterated quality as a result of the deviation enters commerce. Corrective action being taken and its results must be recorded in documents relating to CCP controlling. It is required that each corrective procedure or method being taken and the results should be listed in the QACCP plan (Table 3).

Table 3. Plant's plan table of implementing QACCP system during the food product being processed, transported, stored and sold[a]

CF being affected	Monitoring				Critical limit[c]	Corrective action[b]	Record	Verifying
	Targeted CCP	Measures[b]	Frequency[b]	Personnel				
	Raw material							
	e.g. Variety							
	…							
	Processing							
	e.g. pH							
	…							
	Transporting							
	e.g. Temperature							
	…							
	Storing							
	e.g. Time							
	…							
	Marketing							
	e.g. Temperature							
	…							

[a] After CCP has been determined in Table 2, plant's QACCP plan can be filled in this table and implemented.

[b] These can be established according to knowledge, experience, laboratory-or plant-scale experiments, etc.

[c] The critical limit is the maximum or minimum value that must be controlled of the operating parameters at the critical control point to assure that the critical factor levels of the food product is equal to or better than the acceptable levels appointed by the standards. The critical limit can be established according to knowledge, experience, laboratory-or plant-scale experiments, etc.

2.7. Creation of Procedures for Verifying the QACCP System

Plants have to ensure that they do what they were designed to do. It is necessary that the QACCP plan established is validated that the plants can be successful in ensuring the production of good quality food product. Excepting monitoring procedures, verifying measures needing to be established involve the appointment of auditing methods or procedures, estimating or analyzing methods, personnel management and other validated evaluating methods. The QACCP plans established should be reviewed for conformance with the final rule. The verification procedures appointed may include such activities as review of QACCP plans, CCP records, critical limits and the critical factor sampling and analysis. The QACCP plan including verification tasks can be performed by plant personnel. The process of finding evidence for the accuracy of the QACCP system (e.g. scientific evidence for critical limitations) can also be included in verification tasks. By doing so, the adequacy of QACCP plan established that is, working as intended can be ensured.

2.8. Creation of Procedures for Keeping Record or Memorandum Document

It is necessary that plants maintain certain documents, including its quality analysis and written QACCP plan, and records documenting the monitoring of critical control points, critical limits, verification activities, corrective actions being taken and the handling of processing, transporting, storing and selling deviations. A standard procedure and the requirements for qualified personnel for doing this must be appointed. All recording materials or memorandum documents must be bound for referring later on.

3. Perspective Remarks

The firstly published QACCP that is introduced above is a new concept of controlling food products during their processing, storing and selling in a market. The 8 principles of QACCP introduced in the main context should be quite complete though further minor revision may not be ruled out. By strictly following the 8 principles of QACCP described above, it is reasonable to believe that any food product processed can at least meet its standard previously established.

It is believed that the tables provided are well designed, quite accurate and easily operative. They should be very helpful for putting the 8 principles of QACCP into practice, establishing or implementing a QACCP system being suitable for the quality control of a particular food product being processed. These tables can be employed as working tables in plant operation.

Application of the QACCP system established by following the 8 principles to food manufacturing may be able to achieve the goal of producing a food product with the highest food quality and the quality controlling cost as low as possible. This would attract food manufacturers.

The 8 principles of QACCP may also be helpful for researchers who are going to develop a new food product and a food production line, engineers who are going to design and establish a food processing plant and food plant managers who are going to establish quality control programs. Furthermore, they may be suitable for being included in a text book that is employed in educating students who are studying food quality control.

References

Alli, I. (2004). *Food Quality Assurance: Principles and Practices.* Boca Raton, Florida, USA: CRC Press.

Bauman, H.E. (1995). The origin and concept of HACCP. In A.M. Pearson, and T.R. Duston (Eds.), *HACCP in meat, poultry and fish processing, Advances in Meat Research Series – Volume 10*. London, UK: Blackie Academic and Professional.

FAO. (1999). *The importance of food quality and safety for developing countries.* Rome: Committee on World Food Security, 25th session.

FAO/WHO. (2007). *FAO/WHO guidance to governments on the application of HACCP in small and/or less-developed food businesses* (PDF). ftp://ftp.fao.org/docrep/fao/009/a0799e/a0799e00.

FAO/WHO. (1989a). Codex Standard 174-1989, Codex General Standard for Vegetable Protein Products (VPP). http://www.codexalimentarius.net/web/standard_list.do?lang=en

FAO/WHO. (1989b). Codex Standard 175-1989, Standard for Soy Protein Products. http://www.codexalimentarius.net/web/standard_list.do?lang=en

Kramer, A., and Twigg, B. (1968). Measure of frozen food quality and quality changes. In D. K. Tressler (Ed.), *The freezing preservation of foods, Vol.2* (4th edition,). Westport Conn.: AVIpub. Co.

Mayes, T., and Mortimore, S.E. (2001). *Making the Most of HACCP.* Cambridge, UK: Woodhead Publishing.

Mortimore, S., and Emond, S. (2008). Principles of HACCP: The Importance of HACCP Systems in Food Manufacturing. In G. S. Tucker (Ed.), *Food Biodeterioration and Preservation*. Oxford, UK: Blackwell Publishing Ltd. doi:10.1002/9780470697849.ch2

Otsuka, K., Chuma, H., and Hayami, Y. (1992). Land and Labor Contracts in Agrarian Economies: Theories and Facts. *Journal of Economic Literature 30*, 1965-2018.

Peariso, D. (2006). *Preventing Foreign Material Contamination of Foods.* Pp.52-56. Iowa, USA: Blackwell Publishing.

Prendergast, C. (1999). The Provision of Incentives in Firms. *Journal of Economic Literature 37*, 7-63.

United States Food and Drug Administration. (2007). "Fish and fisheries products hazards and controls guidance, third edition". http://www.cfsan.fda.gov/~comm/haccp4.html.

Unnevehr, L., and Jensen, H.H. (1999). The economic implications of using HACCP as a food safety regulatory standard. *Food Policy*, 24 (6), 625-635.

Vasconcellos, A. (2004). *Quality assurance for the food industry: a practical approach*. Boca Raton, Florida, USA: CRC Press.

Wallace, C., and William, T. (2001). Pre-requisites: a help or a hindrance to HACCP? *Food Control 12*, 235-240.

In: Processed Foods
Editors: Chloe M. Gagne and Daniel B. Jones

ISBN: 978-1-62808-688-1

Chapter 10

ENGINEERED COLLOIDAL SYSTEMS FROM FOOD-GRADE MATERIALS FOR APPLICATIONS IN NOVEL PROCESSED FOODS

Ashok R. Patel

Vandemoortele Centre for Lipid Science and Technology, Laboratory of Food Technology and Engineering, Faculty of Bioscience Engineering, Department of Food Safety and Food Quality, University of Gent, Gent, Belgium

ABSTRACT

Though the novel processed foods have been a topic of global interest in the food and nutrition industries for years, the general consumers seemed incognizant of the obvious benefits that these novel foods could offer. However, recently with the amount of information available in the public domain, modern consumers are becoming more health conscious about their food choices and consequently the market for high-quality novel food products is constantly evolving. As a result, the demands on the improvement of existing food products and the development of novel ones are increasing to such an extent that food industries globally are feeling the necessity to invest heavily into inter-disciplinary research. One of the most important factors to be considered while working in the foods domain, is the choice of raw materials. The strict regulatory requirements in foods means that the starting materials used for the fabrication of novel formulation and delivery systems are food-grade and edible. This chapter gives an account of two such food-grade materials (zein, a hydrophobic plant protein) and (shellac, a hydrophobic resin), which have been extensively utilized as materials of choice to fabricate different types of colloidal delivery systems. Specifically, the chapter focuses on an overview of a range of engineered colloidal systems such as colloidal particles, responsive emulsions and microcapsules prepared using zein and shellac.

INTRODUCTION

The popularity of novel processed food products containing functional ingredients which could provide certain health benefits has been on a rise since the last decade or so. These novel food products (also known as functional foods) are fortified with micronutirents (such as vitamins and minerals) and nutraceuticals (natural bioactives) to provide additional health benefits that go beyond the basic nutrition and nourishment provided by conventional food products. The functional ingredients present in these food products are responsible for influencing specific functions in the body and thereby offering additional benefits for health, well-being, and/or performance beyond their regular nutritional value.

Thus, it is of utmost importance to formulate, stabilize and deliver these functional ingredients in their biologically active forms to develop these novel food products. However, most of these ingredients present multiple problems including low solubility in aqueous medium, sensitivity to degradation in product or *in-vivo* conditions, susceptibility to chemical interactions with food product matrix and most importantly, they may also present significant bioavailability issues. Thus, fortification of food products with additional functional ingredients is often complicated and technologically demanding [1,2]. Due to the complex nature of this problem, we need to look at the interdisciplinary research to develop innovative solutions. The past few years have seen an increased emergence of innovative micro/nano technology for formulation and delivery purposes as applied to foods.

One of the most important factors to be considered while working in the foods domain, is the choice of raw materials. The strict regulatory requirements in foods means that the starting materials used for the fabrication of novel formulation and delivery systems are food-grade and edible. However, there are not many food-grade polymers that can be used for developing colloidal systems because most of them are inherently hydrophilic in nature. For e.g. some of the common food-grade components such as proteins like gelatin, whey proteins, caseins, lactoglobulins etc. and polysaccharides like pectin, starches, gums etc are all hydrophilic in nature. Hence, to make colloidal structures out of them we need to either modify them chemically to impart hydrophobicity or introduce an additional processing step of cross linking by chemical (use of glutaraldehyde, transaminase etc.) or physical means (thermal or pH induced crosslinking) [1].

Thus, food-grade materials which are inherently hydrophobic in nature, can be looked upon as suitable candidates for the fabrication of colloidal systems by a simple alteration of their solubility. This chapter gives an account of two such food-grade materials (zein, a hydrophobic plant protein) and (shellac, a hydrophobic resin), which are recently been utilized as materials of choice to fabricate different types of colloidal delivery systems. Precisely, a brief overview of zein and shellac based colloidal delivery systems are briefly discussed. Due to their water resistant nature and slow digestion, both zein and shellac have a long history of use in pharmaceutical field where they are used as coating materials for developing enteric coated dosage forms which are meant to protect labile drug molecules from harsh acidic pH of stomach and release them in the intestine where they can be absorbed in an unchanged form. Both zein and shellac are natural, biodegradable materials and are also approved for food applications. Recently, we explored the possibility of using zein and shellac to generate discrete spherical colloidal particles which could be further used to entrap and deliver problematic phytochemicals in food products. The possibility of entrapping small molecular weight compounds in zein colloidal particles was further explored to load food-grade colors to generate water dispersible, stable colors. Shellac, also showed non-covalent interactions with other food hydrocolloids such as xanthan gum and gelatin. These interactions were exploited to develop novel colloidal systems such as pH responsive emulsions and biopolymeric microcapsules.

ZEIN BASED COLLOIDAL SYSTEMS

Zein, a prolamine protein obtained from corn, is a well-studied biopolymeric material for development of colloidal particles. The presence of high amount of hydrophobic amino acids makes this protein practically insoluble in water. However, this protein can be easily dissolved in hydro-alcoholic mixtures. On diluting the aqueous ethanolic solution (80:20, EtOH:water) of zein with water, we can cause precipitation of zein to form spherical colloidal particles [3]. Since zein is positively charged below its isoelectric point (pH~ 6.2), if the pH of the solution is adjusted to 3.5-4.0, we can generate discrete colloidal particles in a size range of 80-120 nm without the use of any surfactant. The colloidal particles thus formed, have a high positive surface charge (ζ-potential > 30mV) providing electrostatic stabilization. It has been noted that the electrostatic stabilization is compromised in acidic pH and at high ionic strength (as usually encountered during the passage of food through gastro intestinal tract), leading to an immediate aggregation of these colloidal particles. This problem was solved by providing steric stabilization to zein colloidal particles using another protein-sodium caseinate. Due ot its isoelectric point of 4.6, sodium caseinate is strongly negatively charged at neutral pH. Taking advantage of this charge difference, sodium caseinate stabilized zein colloidal particles were prepared by precipitating zein from hydro-alcoholic solution in sodium caseinate solution adjusted to a pH of 7.0. Due to the presence of a steric layer of caseinate, zein colloidal particles showed increased stability to pH and ionic strength usually encountered in *in-vivo* conditions [4]. Zein colloidal particles were further used to encapsulate hydrophobic phytochemicals for stabilization and delivery and also to generate acid stable food-grade green colors.

Curcumin Loaded Zein Colloidal Particles

Curcumin, a phytochemical with various beneficial pharmacological effects in humans, is one of the most researcher natural component. Though, it has been proved to show strong effects in various studies, its commercial applications is severaly limited due to a combination of factors including: limited solubility in water, extensive first pass metabolism and instability in product and body conditions (e.g. photoinstability, instability in alkaline pH and susceptibility to enzymatic degradation). Several colloidal delivery systems including liposomes, microspheres, dendrimers, micelles, hydrogels and solid lipid nanoscale particles have already been designed to address the above mentioned issues. However, most of these studies were carried out using synthetic and non-edible polymers which makes them unserviceable for food applications.

Since curcumin shows good solubility in ethanol, composite colloidal particles of zein and curcumin were developed by simply co-precipitating a hydro-alcoholic solution of zein and curcumin in aqueous medium (Figure 1). Sodium caseinate was included in the precipitating aqueous medium to provide steric stabilization to make these particles suitable for oral delivery. The maximum loading of curcumin in these particles is only limited by the solubility of curcumin in the starting stock solution (EtOH:water, 80:20). The size of the obtained particles (with 20% loading of curcumin per weight of zein) could be controlled down to around 110 nm. The encapsulation of these hydrophobic curcumin led to an increase in its water dispersibility on one hand due to the same size of colloidal particles and on the other hand, due to the entrapment in the polymeric matrix, the physicochemical stability (pH and photo stability) of curcumin was also improved. Zein is also known to have bioadhesive property and accordingly, *in-vitro* tests conducted on mucomimetic hydrocolloid gel and mucin coated Caco2 cells confirmed that the colloidal particles also showed affinity for the mucin layer. Further, due to the hydrophobicity and slow digestion of zein protein, the release of encapsulated polyphenols could be controlled.

Quercetin Loaded Zein Colloidal Particles

Quercetin is a natural flavonol known to possess wide physiological activities in human owing to its strong anti-oxdiant properties. However, most of these activities are difficult to translate into actual clinical settings due to its low oral bioavailability which results from its low aqueous solubility and high susceptibility to degradation in intestinal pH. To overcome this problem, quercetin was loaded into zein colloidal particles resulting in an increase in aqueous dispersibility as well as protection against degradation due to entrapment in the protein matrix [5]. Protocol followed for the preparation of quercetin loaded zein particles was same as the one used for preparation of curcumin loaded particles.

Due to the extremely low water solubility of quercetin (<0.5 μg/ml), large precipitates (average particle diameter >1 μm) were formed when the ethanolic solution of quercetin was added to water. The dispersion of quercetin needle like crystals was however extremely unstable (due to the relatively large particles size = 4768 nm and a low ζ-potential value = -17.4mV), leading to sedimentation on standing. The shape of the precipitates was controlled by varying the proportion of zein in the stock solution. With increasing proportion of zein there was a decrease in the appearance of needle shaped crystals in the specimen, with a complete disappearance at zein:quercetin ratio of 25:1 wt/wt, where after the particles showed spherical shapes (Figure 2) suggesting effective encapsulation at higher proportions of zein.

Based on photostability and anti-oxidant studies, it was confirmed that entrapment of quercetin in colloidal particles led to its enhanced molecular stability to alkaline pH and UV irradiation.

Zein Colloidal Particles Loaded with Food-Grade Colors

Food coloring plays a vital and a determining role in the processing and the manufacturing of food products because the appearance of products is critical for attracting consumers and influencing their food choices. However, factors such as legislative restrictions, limited number of approved colorants and the processing, formulation and stability issues of the natural colorants severely limits the application of food coloring in actual product formats. Hence, finding alternatives to the currently utilized formulation practices, represents an important area of research. Recently, a simple colloidal approach was reported to prepare color blends by co-incorporating colorants with contrasting aqueous solubility profiles in composite colloidal particles [6]. Curcumin and indigocarmine were selected as water insoluble and water soluble food-grade colorants respectively and incorporated in the colloidal particles prepared from food protein-zein. Spherical composite particles obtained by loading of curcumin and indigocarmine (at different ratios) had mean particle size ranging from 76 to 300 nm. The incorporation of colorants in colloidal particles led to the generation of different shade of color in yellow – green – blue range (Figure 3). The encapsulation also led to the stabilization of individual pigments against photodegradation. Such composite colloidal particles could potentially serve as an approach for developing tuneable coloring system for food and nutraceutical applications.

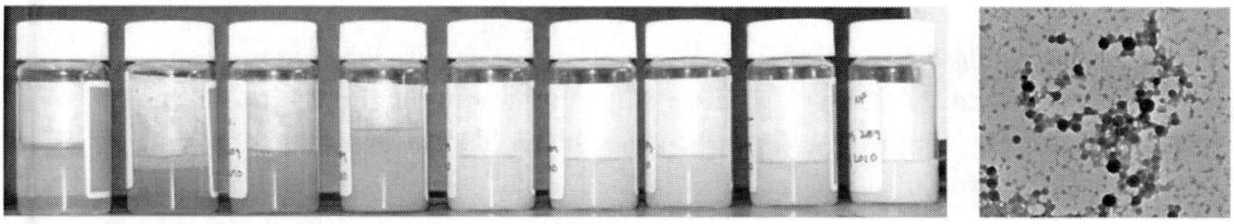

Source: Patel A.R. et al. *Soft Matter*, 2010, 6, 6192-6199.

Figure 1. (On the left) Representative picture of colloidal dispersions prepared using various zein: curcumin (Z: C) ratios in the stock solution of 80: 20 v/v ethanol: water. (From left: pure curcumin; 1: 1; 2: 1; 4: 1; 5: 1; 10: 1; 25: 1; 50: 1 and pure zein). (On the right) TEM image of zein–curcumin composite colloidal particles prepared using zein : curcumin ratio of 5 : 1 in the stock solution (image width = 2 μm).

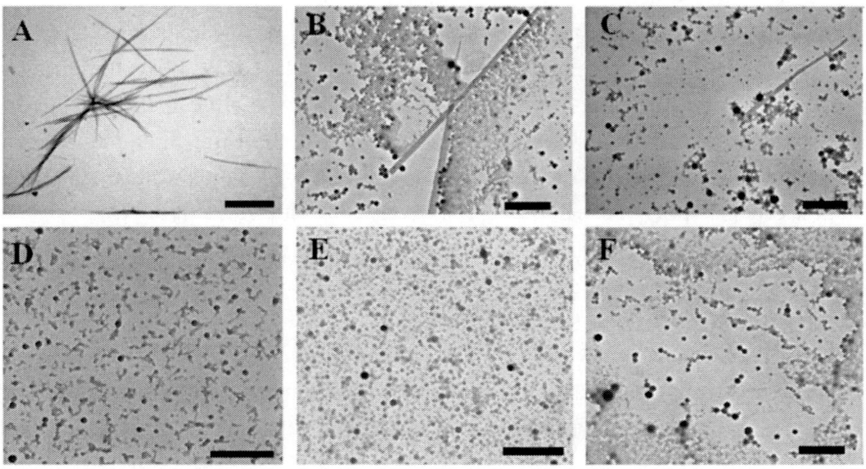

Source: Patel A.R. et al. *Food Chem.*, 2012, 113, 423-429.

Figure 2. TEM images of (A) Needle-like crystals of quercetin obtained by precipitating its alcoholic solution in water (Scale bar = 5 μm); Colloidal particles prepared at different zein:quercetin ratios of (B) 5:1 wt/wt, (C) 10:1 wt/wt; (D) 25:1 wt/wt; (E) 50:1 wt/wt (Scale bar = 1 μm) and (F) blank zein colloidal particles (Scale bar = 1 μm).

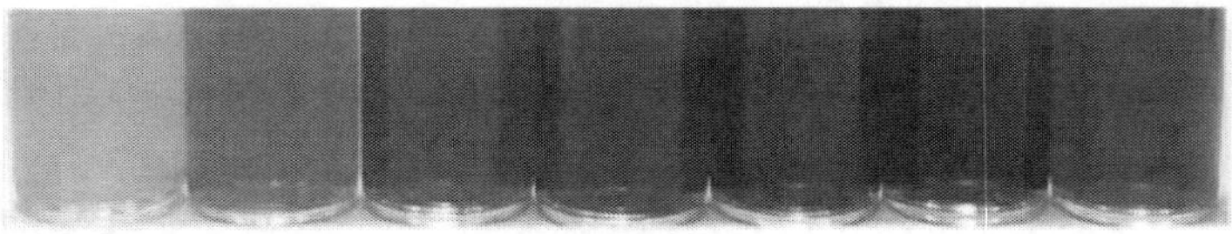

Source: Patel A.R. et al. *Food Chem.*, 2013,141, 1466-1471.

Figure 3. Colloidal dispersions with a tuneable yellow-green-blue color containing (from left to right) zein:Cur:IC (20:1:0; 20:1:0.25; 20:1:0.5; 20:1:1; 20:0.5:1, 20:0.25:1 and 20:0:1). The ratios are expressed on weight basis and the total concentration of zein in all cases was 1 %wt.

SHELLAC BASED COLLOIDAL SYSTEMS

Shellac Based pH Responsive Colloidal Particles

Shellac- a natural, biodegradable resin from insect origin is a complex mixture of polar and non-polar components consisting of polyhydroxy, polycarboxylic esters, lactones and anhydrides. Shellac has a long history of use in the pharmaceutical field and it has also been approved as a food additive (E904). Shellac is acid resistant but shows good solubility in the alkaline pH (pH > 7.0). In pH < 7, it is practically insoluble in water and hence is a very good starting material to form insoluble but dispersed colloidal particles.

The hydrophobic nature of shellac was exploited to generate colloidal particles (size range: 150-300nm) by precipitating shellac from a molecular ethanol solution into an aqueous medium containing a steric stabilizer (xanthan gum). Colloidal particles also showed the pH dependent solubility like shellac wherein the colloidal particles were insoluble in acidic pH but showed solubilization in neutral-alkaline pH of intestine [7]. Such pH responsive particles can be used to develop controlled delivery systems for phytochemicals where a targeted release is required. The application of these pH responsive colloidal particles was further evaluated by encapsulating a model phytochemical-silibinin (a flavolignan with strong liver protective activity). Silibinin shows a low oral bioavailability owing to its degradation in the acidic pH of stomach. Based on *in-vitro* studies, we observed that the encapsulated silibinin in shellac colloidal particles was protected from degradation in the acidic pH of stomach whereas, a burst release was seen when pH was changed to neutral-alkaline pH (Figure 4).

Shellac Based pH Responsive Emulsions

Food-grade polysaccharides tend to show strong interactions with small molecular weight compounds such as polyphenols (epigallocatechin gallate, tannic acid) and resin (shellac) [8-11]. We have shown that these spontaneous interactions can be exploited to generate colloidal complexes which are surface active. These complexes can then be used to stabilize both air-water as well as oil-water interfaces [9,10]. This effect of colloidal complexes is particularly interesting from an application point of view as both air-water as well as oil-water interfaces are most ubiquitously found in most processed food systems. Similarly, colloidal complexes of xanthan gum (polysaccharide) with a resin (shellac) resulted in the enhancement of emulsifying properties of polysaccharide [11]. This was a significant effect considering the fact that xanthan gum is inherently non-surface active. These stabilized emulsions further showed a pH responsive behavior wherein the emulsions could be reversibly switched between stable and flocculated state (Figure 5). Such emulsions could find interesting applications for controlling delivery of functional ingredients.

Shellac Based Microcapsules

Microencapsulation is an enabling technique which have generated huge academic research interest as well as proven industrial applications. On one hand, encapsulation can be used to stabilize and deliver functional ingredients (such as phytochemicals and micronutrients like vitamins and mineral) in fortified food products and on the other hand, encapsulation of flavors and colors can also be used to improve the aesthetics and organoleptic properties of food products in order to make them more acceptable to potential consumers. Hence, there is a great deal of interest in finding food-grade materials which could be used to fabricate novel microcapsules.

Interestingly, non-covalent interactions of food-grade polymers with small molecular weight compounds can be utilized to generate novel microcapsules with different morphologies [12]. Interactions of gelatin with shellac which leads to the formation of colloidal complexes, can also be used to form microstructure by simply extruding gelatin-shellac mixture into an acidic medium [13]. Gelatin-shellac interactions are mediated by a combination of weak interactions (including hydrogen bonding and ionic as well as hydrophobic interactions) as confirmed from the results of isothermal thermal titration. The acid-resistant nature of shellac ensures that the gelatin-shellac complex in the extruded drops solidify into spherical microcapsules with a homogenous internal structure (Figure 6). These novel microcapsules were further evaluated for various encapsulation applications and were found to have interesting properties. Water soluble and insoluble colors, flavors, enzymes and bioactives were successfully encapsulated in these novel structures.

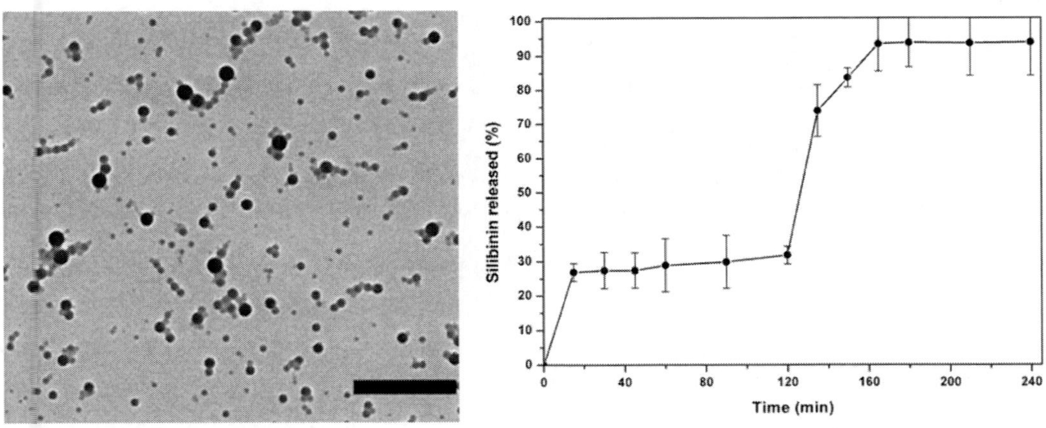

Source: Patel A.R. et al. *Soft Matter*, 2011, 7, 8549-8555.

Figure 4. (On the left) TEM image of shellac colloidal particles loaded with silibinin (Scale bar = 1 μm); (On the right) Release profile of silibinin from colloidal particles in gastric pH (1.2) for initial 2 hours followed by incubation in intestinal pH (7.4) for additional 2 hours. Values reported as mean ± SD 2011, 7, 8549-8555.

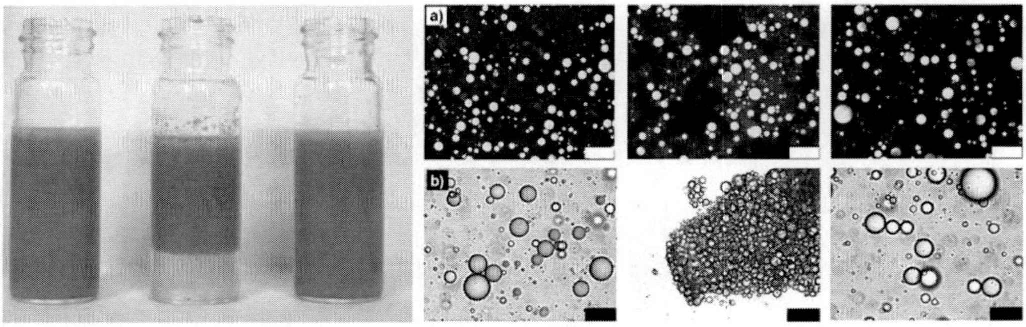

Source: Patel, A.R. et al. *Soft Matter*, 2013, DOI: 10.1039/c3sm27900f.

Figure 5. (On the left) Photographs of emulsions stabilized by xanthan gum (fluorescently labelled) and shellac showing reversible flocculation when pH is changed from neutral to acidic and back to neutral. (On the right) Corresponding CLSM (a) and optical microscopy images (b) of emulsions (scale bars = 50 μm).

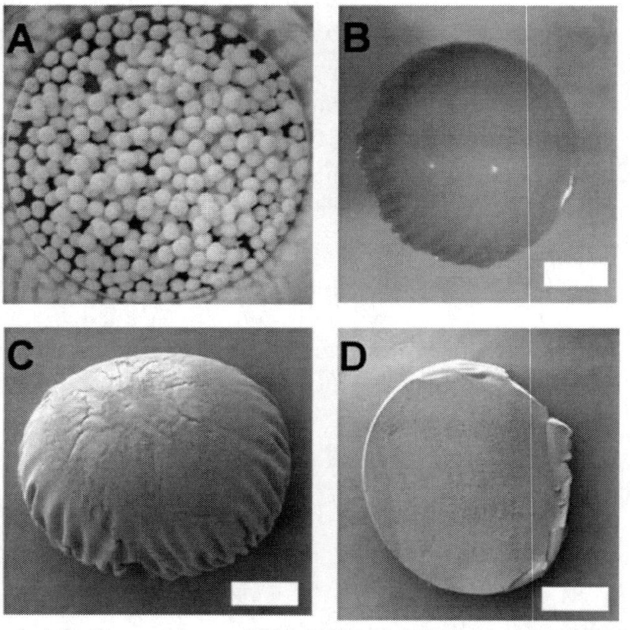

Source: Patel, A.R. et al. *Adv. Funct. Mater.*, 2013, DOI: 10.1002/adfm.201300320.

Figure 6. A) Photograph of microcapsules after minutes of formation; B) Optical microcopy images of an intact microcapsules (scale bars = 100 μm); C, D) SEM images of intact and cross sectioned microcapsule (scale bars = 100 μm).

CONCLUSION

In conclusion, two food-grade materials zein and shellac obtained from natural sources can be effectively utilized to generate novel structures in nano/colloidal and micro size ranges. Zein and shellac based colloidal systems such as nanoparticles and microcapsules can have important future applications especially in formulating problematic bioactives and ensuring their effective *in-vivo* delivery to overcome bioavailability issues. Further, colloidal complexes generated based on the interactions of polysaccharide and shellac can also lead to significant enhancement in the interfacial stabilization properties thus, enabling development of food colloids without the use of emulsifiers. On the other hand, microstructures such as microcapsules developed based on the mutual interactions of biopolymer (gelatin) and shellac have shown good potentials for microencapsulation related applications. These novel structures could easily help us tackle various formulation and delivery issues encountered during the design and development of processed functional foods. In addition, the possibility of creating colloidal dispersions of food-grade colors as well as the encapsulation of aroma and colors in microparticles can be exploited to improve the aesthetics and organoleptic properties of novel food products.

REFERENCES

[1] Patel, A. R. & Velikov, K. P. (2011). Colloidal delivery systems in foods: A general comparision with oral drug delivery, *LWT Food Sci. Tech., 44*, 1958-1964.

[2] Patel, A. R. (2012). Colloidal delivery systems for phytochemicals, *In Nanomedicine and Drug Delivery*, Apple Press Publications, Canada, 17-32.

[3] Patel, A. R., Hu, Y., Tiwari, J. K. & Velikov, K. P. (2010). Synthesis and characterization of zein-curcumin colloidal particles, *Soft Matter*, *6*, 6192-1699.

[4] Patel, A. R., Bouwens, E. C. B. & Velikov, K. P. (2010). Sodium caseinate stabilized zein colloidal particles, *J. Agric. Food Chem., 58*, 12497-12503.

[5] Patel, A. R., Heussen, P. C. M., Hazekamp, J., Drost, E. & Velikov, K. P. (2012). Quercetin loaded biopolymeric colloidal particles prepared by simultaneous precipitation of quercetin with hydrophobic protein in aqueous medium, *Food Chem., 113*, 423-429.

[6] Patel, A. R., Heussen, P. C. M., Drost, E., Hazekamp, J. & Velikov, K. P. Colloidal approach to prepare color blends from colorants with different solubility profiles, *Food Chem., 141*, 1466-1471.

[7] Patel, A. R., Heussen, P. C. M., Hazekamp, J. & Velikov, K. P. (2011). Stabilisation and controlled release of silibinin from pH responsive shellac colloidal particles, *Soft Matter, 7*, 8549-8555.

[8] Patel, A. R., Seijen-ten-Hoorn, J. & Velikov, K. P. (2011). Colloidal complexes from associated water soluble cellulose derivative (methylcellulose) and green tea polyphenol (Epigallocatechin gallate), *J. Colloid. Interf. Sci., 364*, 317-323.

[9] Patel, A. R., Seijen-ten-Hoorn, J., Hazekamp, J., Blijdenstein, T. B. J. & Velikov, K. P. (2013). Colloidal complexation of a macromolecule with a small molecular weight natural polyphenol: implications in modulating polymer functionalities, *Soft Matter, 9*, 1428-1436.

[10] Patel, A. R., Drost, E., Blijdenstein, T. B. J. & Velikov, K. P. (2012). Stable and temperature-responsive surfactant-free foamulsions with high oil-volume fraction, *ChemPhysChem, 13*, 3777-3781.

[11] Patel, A. R., Drost, E., Seijen-ten-Hoorn, J. & Velikov, K. P. Fabrication and characterization of emulsions with pH responsive switchable behavior, *Soft Matter,* DOI:10.1039/C3SM27900F.

[12] Patel, A. R., Nisjje, J. & Velikov, K. P. (2011). Novel polymer-polyphenol beads for encapsulation and microreactor applications, *Soft Matter, 7*, 4294-4301.

[13] Patel, A. R., Chung, D. & Lee, H. G. Novel all-natural microcapsules from gelatin and shellac for biorelated applications, *Adv. Funct. Mater.,* DOI: 10.1002/adfm.201300320.

INDEX

A

D

E

F

G

H

I

J

K

L

M

N

O

P

T

U

V

W

X

Y

Z